ASE Test Preparation Series

Automobile Test

Brakes (Test A5)

4th Edition

THOMSON
DELMAR LEARNING

Australia Canada Mexico Singapore Spain United Kingdom United States

THOMSON
DELMAR LEARNING

Thomson Delmar Learning's ASE Test Preparation Series

Automobile Test for Brakes (Test A5), 4th Edition

Vice President, Technology Professional Business Unit:
Gregory L. Clayton

Product Development Manager:
Kristen L. Davis

Product Manager:
Kimberley Blakey

Editorial Assistant:
Vanessa Carlson

Director of Marketing:
Beth A. Lutz

Marketing Specialist:
Brian McGrath

Marketing Coordinator:
Marissa Maiella

Production Manager:
Andrew Crouth

Production Editor:
Kara A. DiCaterino

Senior Project Editor:
Christopher Chien

XML Architect:
Jean Kaplansky

Cover Design:
Michael Egan

Cover Images:
Portion courtesy of DaimlerChrysler Corporation

Printed in the United States of America
2 3 4 5 XX 11 10 09 08 07

For more information contact Thomson Delmar Learning
Executive Woods
5 Maxwell Drive, PO Box 8007,
Clifton Park, NY 12065-8007
Or find us on the World Wide Web
at www.delmarlearning.com
or www.trainingbay.com

ISBN-10: 1-4180-3882-2
ISBN-13: 978-1-4180-3882-3

NOTICE TO THE READER

Publisher does not warrant or guarantee any of the products described herein or perform any independent analysis in connection with any of the product information contained herein. Publisher does not assume, and expressly disclaims, any obligation to obtain and include information other than that provided to it by the manufacturer.

The reader is expressly warned to consider and adopt all safety precautions that might be indicated by the activities herein and to avoid all potential hazards. By following the instructions contained herein, the reader willingly assumes all risks in connection with such instructions.

The publisher makes no representation or warranties of any kind, including but not limited to, the warranties of fitness for particular purpose or merchantability, nor are any such representations implied with respect to the material set forth herein, and the publisher takes no responsibility with respect to such material. The publisher shall not be liable for any special, consequential, or exemplary damages resulting, in whole or part, from the readers' use of, or reliance upon, this material.

Contents

Section 5 Sample Test for Practice

Section 6 Additional Test Questions for Practice

Section 7 Appendices

Preface

Delmar Learning is very pleased that you have chosen our ASE Test Preparation Series to prepare yourself for the automotive ASE Examination. These guides are available for all of the automotive areas including A1–A8, the L1 Advanced Diagnostic Certification, the P2 Parts Specialist, the C1 Service Consultant and the X1 Undercar Specialist. These guides are designed to introduce you to the Task List for the test you are preparing to take, give you an understanding of what you are expected to be able to do in each task, and take you through sample test questions formatted in the same way the ASE tests are structured.

If you have a basic working knowledge of the discipline you are testing for, you will find Delmar Learning's ASE Test Preparation Series to be an excellent way to understand the "must know" items to pass the test. These books are not textbooks. Their objective is to prepare the technician who has the requisite experience and schooling to challenge ASE testing. It cannot replace the hands-on experience or the theoretical knowledge required by ASE to master vehicle repair technology. If you are unable to understand more than a few of the questions and their explanations in this book, it could be that you require either more shop-floor experience or further study. Some resources that can assist you with further study are listed on the rear cover of this book.

Each book begins with an item-by-item overview of the ASE Task List with explanations of the minimum knowledge you must possess to answer questions related to the task. Following that there are 2 sets of sample questions followed by an answer key to each test and an explanation of the answers to each question. A few of the questions are not strictly ASE format but were included because they help teach a critical concept that will appear on the test. We suggest that you read the complete Task List Overview before taking the first sample test. After taking the first test, score yourself and read the explanation to any questions that you were not sure about, including the questions you answered correctly. Each test question has a reference back to the related task or tasks that it covers. This will help you to go back and read over any area of the task list that you are having trouble with. Once you are satisfied that you have all of your questions answered from the first sample test, take the additional tests and check them. If you pass these tests, you will be prepared to do well on the ASE test.

Our Commitment to Excellence

The 4th edition of Delmar Learning's ASE Test Preparation Series has been through a major revision with extensive updates to the ASE's task lists, test questions, and answers and explanations. Delmar Learning has sought out the best technicians in the country to help with the updating and revision of each of the books in the series.

About the Series Advisor

To promote consistency throughout the series, a series advisor took on the task of reading, editing, and helping each of our experts give each book the highest level of accuracy possible. Dan Perrin has served in the role of Series Advisor for the 4th edition of the ASE Test Preparation Series. Dan began ASE testing with the first series of tests in 1972 and has been continually certified ever since. He holds ASE master status in automotive, truck, collision, and machinist. He is also L1, L2, and alternated fuels certified, along with some others that have expired. He has been an automotive educator since 1979, having taught at the secondary, post-secondary, and industry levels. His service includes participation on boards that include the North American Council of Automotive Teachers (NACAT), the Automotive Industry Planning Council (AIPC), and the National Automotive Technicians Education Foundation (NATEF). Dan currently serves as the Executive Manager of NACAT and Director of the NACAT Education Foundation.

Thanks for choosing Delmar Learning's ASE Test Preparation Series. All of the writers, editors, Delmar Staff, and myself have worked very hard to make this series second to none. I know you are going to find this book accurate and easy to work with. It is our objective to constantly improve our product at Delmar by responding to feedback.

If you have any questions concerning the books in this series, you can email me at: autoexpert@trainingbay.com.

Dan Perrin
Series Advisor

The History and Purpose of ASE

ASE began as the National Institute for Automotive Service Excellence (NIASE). It was founded as a non-profit independent entity in 1972 by a group of industry leaders with the single goal of providing a means for consumers to distinguish between incompetent and competent technicians. It accomplishes this goal by testing and certification of repair and service professionals. From this beginning it has evolved to be known simply as ASE (Automotive Service Excellence) and today offers more than 40 certification exams in automotive, medium/heavy duty truck, collision, engine machinist, school bus, parts specialist, automobile service consultant, and other industry-related areas. At this time there are more than 400,000 professionals with current ASE certifications. These professionals are employed by new car and truck dealerships, independent garages, fleets, service stations, franchised service facilities, and more. ASE continues its mission by also providing information that helps consumers identify repair facilities that employ certified professionals through its Blue Seal of Excellence Recognition Program. Shops that have a minimum of 75% of their repair technicians ASE certified and meet other criteria can apply for and receive the Blue Seal of Excellence Recognition from ASE.

ASE recognized that educational programs serving the service and repair industry also needed a way to be recognized as having the faculty, facilities, and equipment to provide a quality education to students wanting to become service professionals. Through the combined efforts of ASE, industry, and education leaders, the non-profit National Automotive Technicians Education Foundation (NATEF) was created to evaluate and recognize training programs. Today more than 2000 programs are ASE certified under the standards set by the service industry. ASE/NATEF also has a certification of industry (factory) training program known as CASE. CASE stands for Continuing Automotive Service Education and recognizes training provided by replacement parts manufacturers as well as vehicle manufacturers.

ASE certification testing is administered by the American College Testing (ACT). Strict standards of security and supervision at the test centers insure that the technician who holds the certification earned it. Additionally ASE certification also requires that the person passing the test to be able to demonstrate that they have two years of work experience in the field before they can be certified. Test questions are developed by industry experts that are actually working in the field being tested. There is more detail on how the test is developed and administered in the next section. Paper and pencil tests are administered twice a year at over seven hundred locations in the United States. Computer based testing is now also available with the benefit of instant test results at certain established test centers. The certification is valid for five years and can be recertified by retesting. So that consumers can recognize certified technicians, ASE issues a jacket patch, certificate, and wallet card to certified technicians and makes signs available to facilities that employ ASE certified technicians.

You can contact ASE at any of the following:

National Institute for Automotive Service Excellence
101 Blue Seal Drive S.E.
Suite 101
Leesburg, VA 20175
Telephone 703-669-6600
FAX 703-669-6123
www.ase.com

2 Take and Pass Every ASE Test

Participating in an Automotive Service Excellence (ASE) voluntary certification program gives you a chance to show your customers that you have the "know-how" needed to work on today's modern vehicles. The ASE certification tests allow you to compare your skills and knowledge to the automotive service industry's standards for each specialty area.

If you are the "average" automotive technician taking this test, you are in your mid-thirties and have not attended school for about fifteen years. That means you probably have not taken a test in many years. Some of you, on the other hand, have attended college or taken postsecondary education courses and may be more familiar with taking tests and with test-taking strategies. There is, however, a difference in the ASE test you are preparing to take and the educational tests you may be accustomed to.

How are the tests administered?

ASE test are administered at over 750 test sites in local communities. Paper and pencil tests are the type most widely available to technicians. Each tester is given a booklet containing questions with charts and diagrams where required. You can mark in this test booklet but no information entered in the booklet is scored. Answers are recorded on a separate answer sheet. You will enter your answers, using a number 2 pencil only. ASE recommends you bring four sharpened number 2 pencils that have erasers. Answer choices are recorded by coloring in the blocks on the answer sheet. The answer sheets are scanned electronically and the answers tabulated. For test security, test booklets include randomly generated questions. Your answer key must be matched to the proper booklet so it is important to correctly enter the booklet serial number on the answer sheet. All instructions are printed on the test materials and should be followed carefully.

ASE has introduced Computer Based Testing (CBT) at some locations. While the test content is the same for both testing methods the CBT tests have some unique requirements and advantages. It is strongly recommended that technicians considering the CBT tests go the ASE web page at www.ASE.com and review the conditions and requirements for this type of test. There is a demonstration of a CBT that allows you to experience this type of test before you register. Some technicians find this style of testing provides an advantage, while others find operating the computer a distraction. One significant benefit of CBT is the availability of instant results. You can receive your test results before you leave the test center. CBT testing also offers increased flexibility in scheduling. The cost for taking CBTs is slightly higher than paper and pencil tests and the number of testing sites is limited. The first time test taker may be more comfortable with the paper and pencil tests but technicians now have a choice.

Who Writes the Questions?

The questions are written by service industry experts in the area being tested. Each area will have its own technical experts. Questions are entirely job related. They are designed to test the skills you need to be a successful technician. Theoretical knowledge is important and necessary to answer the questions, but the ability to apply that knowledge is the basis of ASE test questions.

Each question has its roots in an ASE "item-writing" workshop where service representatives from automobile manufacturers (domestic and import), aftermarket parts and equipment manufacturers, working technicians, and vocational educators meet in a workshop setting to share ideas and translate them into test questions. Each test question written by these experts must survive review by all members of the group.

The questions are written to deal with practical application of soft skills and system knowledge experienced by technicians in their day-to-day work.

All questions are pre-tested and quality-checked on a national sample of technicians. Those questions that meet ASE standards of quality and accuracy are included in the scored sections of the tests; the "rejects" are sent back to the drawing board or discarded altogether.

Each certification test is made up of between forty and eighty multiple-choice questions.

Note: Each test could contain additional questions that are included for statistical research purposes only. Your answers to these questions will not affect your score, but since you do not know which ones they are, you should answer all questions on the test. The five-year Recertification Test will cover the same content areas as those listed above. However, the number of questions in each content area of the Recertification Test will be reduced by about one-half.

Objective Tests

A test is called an objective test if the same standards and conditions apply to everyone taking the test and there is only one correct answer to each question.

Objective tests primarily measure your ability to recall information. A well-designed objective test can also test your ability to understand, analyze, interpret, and apply your knowledge. Objective tests include true-false, multiple choice, fill in the blank, and matching questions. ASE's tests consist exclusively of four-part multiple-choice objective questions.

The following are some strategies that may be applied to your tests.

Before beginning to take an objective test, quickly look over the test to determine the number of questions, but do not try to read through all of the questions. In an ASE test, there are usually between forty and eighty questions, depending on the subject. Read through each question before marking your answer. Answer the questions in the order they appear on the test. Leave the questions blank that you are not sure of and move on to the next question. You can return to those unanswered questions after you have finished the others. They may be easier to answer at a later time after your mind has had additional time to consider them on a subconscious level. In addition, you might find information in other questions that will help you recall the answers to some of them.

Do not be obsessed by the apparent pattern of responses. For example, do not be influenced by a pattern like **D, C, B, A, D, C, B, A** on an ASE test.

There is also a lot of folk wisdom about taking objective tests. For example, there are those who would advise you to avoid response options that use certain words such as *all, none, always, never, must,* and *only,* to name a few. This, they claim, is because nothing in life is exclusive. They would advise you to choose response options that use words that allow for some exception, such as *sometimes, frequently, rarely, often, usually, seldom,* and *normally.* They would also advise you to avoid the first and last option (A and D) because test writers, they feel, are more comfortable if they put the correct answer in the middle (B and C) of the choices. Another recommendation often offered is to select the option that is either shorter or longer than the other three choices because it is more likely to be correct. Some would advise you to never change an answer since your first intuition is usually correct.

Although there may be a grain of truth in this folk wisdom, ASE test writers try to avoid them and so should you. There are just as many **A** answers as there are **B** answers, just as many **D** answers as **C** answers. As a matter of fact, ASE tries to balance the answers at about 25 percent per choice **A, B, C,** and **D.** There is no intention to use "tricky" words, such as outlined above. Put no credence in the opposing words "sometimes" and "never," for example.

Multiple-choice tests are sometimes challenging because there are often several choices that may seem possible, and it may be difficult to decide on the correct choice. The best strategy, in this case, is to first determine the correct answer before looking at the options. If you see the answer you decided on, you should still examine the options to make sure that none seem more correct than yours. If you do not know or are not sure of the answer, read each option very carefully and try to eliminate those options that you know to be wrong. That way, you can often arrive at the correct choice through a process of elimination.

If you have gone through all of the test and you still do not know the answer to some of the questions, then guess. Yes, guess. You then have at least a 25 percent chance of being correct. If you leave the question blank, you have no chance. Your score is based on the number of questions answered correctly.

Preparing for the Exam

The main reason we have included so many sample and practice questions in this guide is, simply, to help you learn what you know and what you don't know. We recommend that you work your way through each question in this book. Before doing this, carefully look through Section 3; it contains a description and explanation of the question types you'll find on an ASE exam.

Once you understand what the questions will look like, move to the sample test. Answer one of the sample questions (Section 5) then read the explanation (Section 7) to the answer for that question. If you don't feel you understand the reasoning for the correct answer, go back and read the overview (Section 4) for the task that is related to that question. If you still don't feel you have a solid understanding of the material, identify a good source of information on the topic, such as a textbook, and do some more studying.

After you have completed all of the sample test items and reviewed your answers, move to the additional questions (Section 6). This time answer the questions as if you were taking an actual test. Do not use any reference or allow any interruptions in order to get a feel for how you will do on an actual test. Once you have answered all of the questions, grade your results using the answer key in Section 7. For every question that you gave a wrong answer to, study the explanations to the answers and/or the overview of the related task areas. Try to determine the root cause for your missing the question. The easiest thing to correct is learning the correct technical content. The hardest thing to correct is behaviors that lead you to a wrong conclusion. If you knew the information but still got it wrong there is a behavior problem that will need to be corrected. An example would be reading too quickly and skipping over words that affect your reasoning. If you can identify what you did that caused you to answer the question incorrectly you can eliminate that cause and improve your score. Here are some basic guidelines to follow while preparing for the exam:

- Focus your studies on those areas you are weak in.
- Be honest with yourself while determining if you understand something.
- Study often but in short periods of time.
- Remove yourself from all distractions while studying.
- Keep in mind the goal of studying is not just to pass the exam, the real goal is to learn!
- Prepare physically by getting a good night's rest before the test and eat meals that provide energy but do not cause discomfort.
- Arrive early to the test site to avoid long waits as test candidates check in and to allow all of the time available for your tests.

During the Test

On paper and pencil tests you will be placing your answers on a sheet where you will be required to color in your answer choice. Stray marks or incomplete erasures may be picked up as an answer by the electronic reader, so be sure only your answers end up on the sheet. One of the biggest problems an adult faces in test taking, it seems, is placing the answer in the correct spot on the answer sheet. Make certain that you mark your answer for, say, question 21, in the space on the answer sheet designated for the answer for question 21. A correct response in the wrong line will probably result in two questions being marked wrong, one with two answers (which could include a correct answer but will be scored wrong) and the other with no answer. Remember, the answer sheet on the written test is machine scored and can only "read" what you have colored in.

If you finish answering all of the questions on a test and have remaining time, go back and review the answers to those questions that you were not sure of. You can often catch careless errors by using the remaining time to review your answers. Carefully check your answer sheet for blank answer blocks or missing information.

At practically every test, some technicians will invariably finish ahead of time and turn their papers in long before the final call. Some technicians may be doing recertification tests and others may be taking fewer tests than you. Do not let them distract or intimidate you.

It is not wise to use less than the total amount of time that you are allotted for a test. If there are any doubts, take the time for review. Any product can usually be made better with some additional effort. A test is no exception. It is not necessary to turn in your test paper until you are told to do so.

Testing Time Length

An ASE written test session is four hours. You may attempt from one to a maximum of four tests in one session. It is recommended, however, that no more than a total of 225 questions be attempted at any test session. This will allow for just over one minute for each question.

Visitors are not permitted at any time. If you wish to leave the test room, for any reason, you must first ask permission. If you finish your test early and wish to leave, you are permitted to do so only during specified dismissal periods.

You should monitor your progress and set an arbitrary limit to how much time you will need for each question. This should be based on the number of questions you are attempting. It is suggested that you wear a watch because some facilities may not have a clock visible to all areas of the room.

Computer-Based Tests are allotted a testing time according to the number of questions ranging from one half hour to one and one half hours. Advanced level tests are allowed two hours. This time is by appointment and you should be sure to be on time to insure that you have all of the time allocated. If you arrive late for a CBT test appointment you will only have the amount of time remaining on your appointment.

Your Test Results!

You can gain a better perspective about tests if you know and understand how they are scored. ASE's tests are scored by American College Testing (ACT), a nonpartial, unbiased organization having no vested interest in ASE or in the automotive industry.

Each question carries the same weight as any other question. For example, if there are fifty questions, each is worth 2 percent of the total score. The passing grade is 70 percent. That means you must correctly answer thirty-five of the fifty questions to pass the test.

The test results can tell you:

- where your knowledge equals or exceeds that needed for competent performance, or
- where you might need more preparation.

Your ASE test score report is divided into content areas and will show the number of questions in each content area and how many of your answers were correct. These numbers provide information about your performance in each area of the test. However, because there may be a different number of questions in each content area of the test, a high percentage of correct answers in an area with few questions may not offset a low percentage in an area with many questions.

It should be noted that one does not "fail" an ASE test. The technician who does not pass is simply told "More Preparation Needed." Though large differences in percentages may indicate problem areas, it is important to consider how many questions were asked in each area. Since each test evaluates all phases of the work involved in a service specialty, you should be prepared in each area. A low score in one area could keep you from passing an entire test.

There is no such thing as average. You cannot determine your overall test score by adding the percentages given for each task area and dividing by the number of areas. It doesn't work that way because there generally are not the same number of questions in each task area. A task area with twenty questions, for example, counts more toward your total score than a task area with ten questions.

Your test report should give you a good picture of your results and a better understanding of your strengths and weaknesses for each task area.

If you fail to pass the test, you may take it again at any time it is scheduled to be administered. You are the only one who will receive your test score. Test scores will not be given over the telephone by ASE nor will they be released to anyone without your written permission.

3 Types of Questions on an ASE Exam

ASE certification tests are often thought of as being tricky. They may seem to be tricky if you do not completely understand what is being asked. The following examples will help you recognize certain types of ASE questions and avoid common errors.

Paper-and-pencil tests and computer-based test questions are identical in content and difficulty. Most initial certification tests are made up of forty to eighty multiple-choice questions. Multiple-choice questions are an efficient way to test knowledge. To answer them correctly, you must think about each choice as a possibility, and then choose the one that best answers the question. To do this, read each word of the question carefully. Do not assume you know what the question is about until you have finished reading it.

About 10 percent of the questions on an actual ASE exam will use an illustration. These drawings contain the information needed to correctly answer the question. The illustration must be studied carefully before attempting to answer the question. Often, technicians look at the possible answers then try to match up the answers with the drawing. Always do the opposite; match the drawing to the answers. When the illustration is showing an electrical schematic or another system in detail, look over the system and try to figure out how the system works before you look at the question and the possible answers.

Multiple-Choice Questions

The most common type of question used on ASE Tests is the multiple-choice question. This type of question contains three "distracters" (wrong answers) and one "key" (correct answer). When the questions are written effort is made to make the distracters plausible to draw an inexperienced technician to one of them. This type of question gives a clear indication of the technician's knowledge. Using multiple criteria including cross-sections by age, race, and other background information, ASE is able to guarantee that a question does not bias for or against any particular group. A question that shows bias toward any particular group is discarded. If you encounter a question that you are unsure of, reverse engineer it by eliminating the items that it cannot be. For example:

A rocker panel is a structural member of which vehicle construction type?

A. Front-wheel drive
B. Pickup truck
C. Unibody
D. Full-frame

Analysis:

This question asks for a specific answer. By carefully reading the question, you will find that it asks for a construction type that uses the rocker panel as a structural part of the vehicle.

Answer A is wrong. Front-wheel drive is not a vehicle construction type.
Answer B is wrong. A pickup truck is not a type of vehicle construction.
Answer C is correct. Unibody design creates structural integrity

by welding parts together, such as the rocker panels, but does not require exterior cosmetic panels installed for full strength.
Answer D is wrong. Full-frame describes a body-over-frame construction type that relies on the frame assembly for structural integrity.

Therefore, the correct answer is C. If the question was read quickly and the words "construction type" were passed over, answer A may have been selected.

EXCEPT Questions

Another type of question used on ASE tests has answers that are all correct except one. The correct answer for this type of question is the answer that is wrong. The word "**EXCEPT**" will always be in capital letters. You must identify which of the choices is the wrong answer. If you read quickly through the question, you may overlook what the question is asking and answer the question with the first correct statement. This will make your answer wrong. An example of this type of question and the analysis is as follows:

All of the following are tools for the analysis of structural damage **EXCEPT:**

A. height gauge
B. tape measure.
C. dial indicator.
D. tram gauge.

Analysis:

The question really requires you to identify the tool that is not used for analyzing structural damage. All tools given in the choices are used for analyzing structural damage except one. This question presents two basic problems for the test-taker who reads through the question too quickly. It may be possible to read over the word "**EXCEPT**" in the question or not think about which type of damage analysis would use answer C. In either case, the correct answer may not be selected. To correctly answer this question, you should know what tools are used for the analysis of structural damage. If you cannot immediately recognize the incorrect tool, you should be able to identify it by analyzing the other choices.

Answer A is wrong. A height gauge may be used to analyze structural damage.
Answer B is wrong. A tape measure may be used to analyze structural damage.
Answer C is correct. A dial indicator may be used as a damage analysis tool for moving parts, such as wheels, wheel hubs, and axle shafts, but would not be used to measure structural damage.
Answer D is wrong. A tram gauge is used to measure structural damage.

Technician A, Technician B Questions

The type of question that is most popularly associated with an ASE test is the "Technician A says . . . Technician B says . . . Who is right?" type. In this type of question, you must identify the correct statement or statements. To answer this type of question correctly, you must carefully read each technician's statement and judge it on its own merit to determine if the statement is true.

Sometimes this type of question begins with a statement about some analysis or repair procedure. This is often referred to as the stem of the question and provides the setup or background information required to understand the conditions the question is based on. This is followed by two statements about the cause of the concern, proper inspection, identification, or repair choices. You are asked whether the first statement, the second statement, both statements, or neither statement is correct. Analyzing this type of question is a little easier than the other types because there are only two ideas to consider although there are still four choices for an answer.

Technician A, Technician B questions are really double true or false questions. The best way to analyze this kind of question is to consider each technician's statement separately. Ask yourself, is A true or false? Is B true or false? Then select your answer from the four choices. An important point to remember is that an ASE Technician A, Technician B question will never have Technician A and B directly disagreeing with each other. That is why you must evaluate each statement independently.

An example of this type of question and the analysis of it follows.

A vehicle comes into the shop with a gas gauge that will not register above one half full. When the sending unit circuit is disconnected the gauge reads empty and when it is connected to ground the gauge goes to full. Technician A says that the sending unit is shorted to ground. Technician B says the gauge circuit is working and the sending unit is likely the problem. Who is right?

A. A only
B. B only
C. Both A and B
D. Neither A nor B

Analysis:

Reading of the stem of the question sets the conditions of the customer concern and establishes what information is gained from testing. General knowledge of gauge circuits and test procedures are needed to correctly evaluate the technician's conclusions. Note: Avoid being distracted by experience with unusual or problem vehicles that you may have worked on, Other technicians taking the same test do not have that knowledge, so it should not be used as the basis of your answers.

Technician A is wrong because a shorted to ground sending unit would produce a gauge reading equivalent to the test conditions of a grounding the circuit and produce a full reading.
Technician B is correct because the gauge spans when going from an open circuit to a completely
grounded circuit. This would tend to indicate that the problem had to be in the sending unit.
Answer C is not correct. Both technicians are identifying the problem as a sending unit but technician A qualified the problem as a specific type of failure (grounded) that would not have caused the symptoms of the vehicle.
Answer D is not correct because technician B's diagnosis is a possible cause of the conditions identified.

Most-Likely Questions

Most-Likely questions are somewhat difficult because only one choice is correct while the other three choices are nearly correct. An example of a Most-Likely-cause question is as follows:

The Most-Likely cause of reduced turbocharger boost pressure may be a:

A. wastegate valve stuck closed.
B. wastegate valve stuck open.
C. leaking wastegate diaphragm.
D. disconnected wastegate linkage.

Analysis:

Answer A is wrong. A wastegate valve stuck closed increases turbocharger boost pressure.
Answer B is correct. A wastegate valve stuck open decreases turbocharger boost pressure.
Answer C is wrong. A leaking wastegate valve diaphragm increases turbocharger boost pressure.

Answer D is wrong. A disconnected wastegate valve linkage will increase turbocharger boost pressure.

LEAST-Likely Questions

Notice that in Most-Likely questions there is no capitalization. This is not so with LEAST-Likely type questions. For this type of question, look for the choice that would be the LEAST-Likely cause of the described situation. Read the entire question carefully before choosing your answer. An example is as follows:

What is the LEAST-Likely cause of a bent pushrod?

A. Excessive engine speed
B. A sticking valve
C. Excessive valve guide clearance
D. A worn rocker arm stud

Analysis:

Answer A is wrong. Excessive engine speed may cause a bent pushrod.
Answer B is wrong. A sticking valve may cause a bent pushrod.
Answer C is correct. Excessive valve clearance will not generally cause a bent pushrod.
Answer D is wrong. A worn rocker arm stud may cause a bent pushrod.
You should avoid relating questions to those unusual situations that you may have encountered and answer based on the technical and mechanical possibilities.

Summary

There are no four-part multiple-choice ASE questions having "none of the above" or "all of the above" choices. ASE does not use other types of questions, such as fill-in-the-blank, completion, true-false, word-matching, or essay. ASE does not require you to draw diagrams or sketches. If a formula or chart is required to answer a question, it is provided for you. There are no ASE questions that require you to use a pocket calculator.

4 Overview of the Task List

Brakes (Test A5)

The following section includes the task areas and task lists for this test and a written overview of the topics covered in the test.

The task list describes the actual work you should be able to do as a technician that you will be tested on by the ASE. This is your key to the test and you should review this section carefully. We have based our sample test and additional questions upon these tasks, and the overview section will also support your understanding of the task list. ASE advises that the questions on the test may not equal the number of tasks listed; the task lists tell you what ASE expects you to know how to do and be ready to be tested upon.

At the end of each question in the Sample Test and Additional Test Questions sections, a letter and number will be used as a reference back to this section for additional study. Note the following example: B.5.

B. Drum Brake Diagnosis and Repair (5 Questions)

Task B.5 Using proper safety procedures, remove, clean, and inspect brake shoes/linings, springs, pins, self-adjusters, levers, clips, brake backing (support) plates, and other related brake hardware; determine needed repairs.

Example:

1. Which of the following should be used to clean the bores on wheel cylinders?
   A. Cleaning solvent
   B. Oil soap
   C. Hot water
   D. Denatured alcohol (B.5)

Analysis:

Question #1

Answer A is wrong. Cleaning solvent will leave a residue on the bores that is undesirable.

Answer B is wrong. Oil soap will leave a petroleum film on the bore that will cause the piston seals to swell.

Answer C is wrong. Using hot water will not clean the bores well enough and also would promote rust.

Answer D is correct. The bores of a wheel cylinder should only be cleaned with denatured or isopropyl alcohol. Other cleaning agents may leave a film behind, which can contaminate the fluid

Task List and Overview

A. Hydraulic System Diagnosis and Repair (12 Questions)

Task A.1 Master Cylinder (3 Questions)

Task A.1.1 Diagnose poor stopping, dragging, high or low pedal, hard or spongy pedal caused by problems in the master cylinder; determine needed repairs.

Modern brake systems consist of front disc brakes, rear drum brakes, master cylinder, lines, valves, power-assist booster, brake pedal, and flexible hoses. Many vehicles today are equipped with front disc and rear disc brakes. For safe and effective brake operation, all components in the brake system must be in proper working order.

The heart of the brake system is the **master cylinder.** The master cylinder is a hydraulic pump that is operated by a pushrod connected to the brake pedal or assisted by a brake booster. The purpose of the master cylinder is to convert the driver's mechanical force into hydraulic pressure. Master cylinders are divided into two separate hydraulic systems. On most RWD vehicles, one system controls the rear brakes and the other system the front brakes. If a hydraulic failure occurs in one system of a dual-system master cylinder, the other system remains operational. This provides a greater protection against a total hydraulic brake failure.

Some master cylinders are split diagonally. In this system, one-half of the master controls one front wheel on one side and one rear wheel on the other side. The diagonal split system is most often found on front-wheel drive vehicles. Master cylinders are manufactured aluminum and have a separate plastic reservoir. Aluminum master cylinders are anodized, which is intended to extend the life of the master cylinder by reducing corrosion.

Some common symptoms of master cylinder-related problems are: an improperly adjusted brake light switch or cruise control switch causing brake drag by keeping the brake pedal partially depressed. This causes the master cylinder to maintain pressure on the brake system. Also, a binding brake pedal caused by worn pivot bushings can cause brake drag.

Air in the hydraulic system will cause a low, spongy pedal. An internal leak in the master cylinder cups may also cause a low, sinking pedal. Always consider the level and condition of the brake fluid when diagnosing master cylinder problems. Refer to Task A.4.1 and A.4.4 for more on brake fluid. A worn or corroded internal master cylinder bore may cause excessive pedal effort and brake drag. Another cause for brake drag is a blocked master cylinder compensation port or vent port.

Brake drag will cause overheated brake linings. It is important to inspect the entire brake system after determining a master cylinder problem.

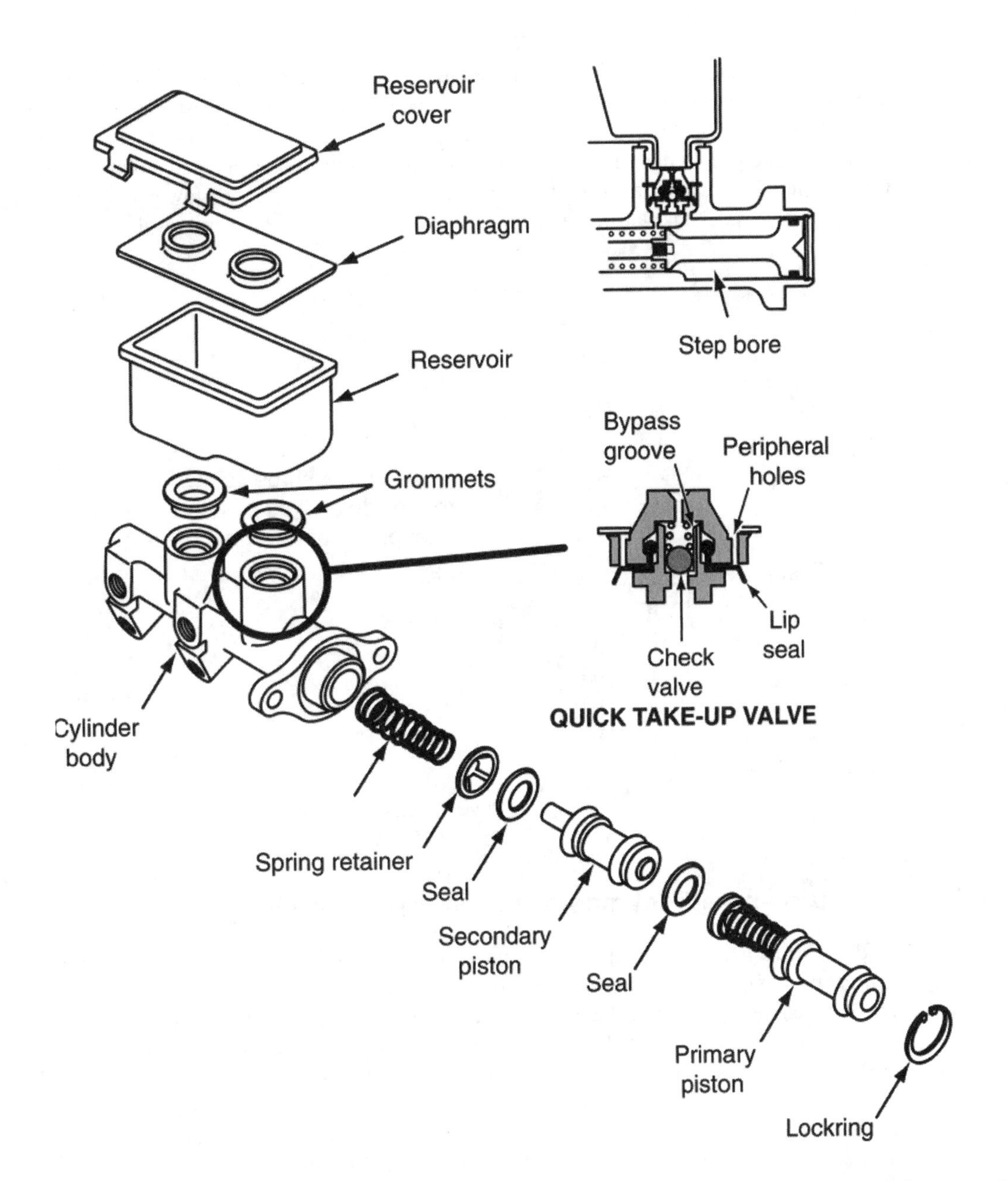

Task A.1.2 Diagnose problems in the step bore master cylinder and internal valves (e.g., volume control devices, quick take-up valve, fast-fill valve, pressure-regulating valve); determine needed repairs.

The **step bore,** or **quick take-up,** master cylinder incorporates two different size bore diameters. The step bore master cylinder improves brake effectiveness in low drag disc brake calipers, which improves fuel economy. In a step bore master cylinder, the forward bore is slightly smaller than the rear bore. This design allows the small bore of the cylinder to supply high pressure with low volume. The large bore supplies high volume at low pressure. The large bore system initially supplies a large volume of fluid to move the caliper piston and apply the brake pads. Once the pads are in contact with the brake rotors, the small bore applies higher hydraulic pressure.

The aluminum bore of a step bore master cylinder is anodized and cannot be honed. Honing would remove the hard bore coating. If the bore is pitted or damaged, the master cylinder must be replaced.

Step bore master cylinders must be bench bled due to severe mounting angle, the quick take-up valve, and the four-port step bore design. It is nearly impossible to bleed all the air from the master cylinder once it is installed.

Diagnosing problems with a step bore master cylinder requires attention to details that are not common to other types of master cylinders. Never overfill a step bore master cylinder. Overfilling can cause the front brakes to drag or cause excessive brake pedal effort. Another cause for excessive pedal effort or a hard pedal on a step bore master cylinder is a clogged or plugged quick take-up valve. On the other hand, if a quick take-up valve opens too early, this can cause a low pedal condition.

A defective quick take-up valve could cause a front caliper to drag and result in premature, uneven pad wear.

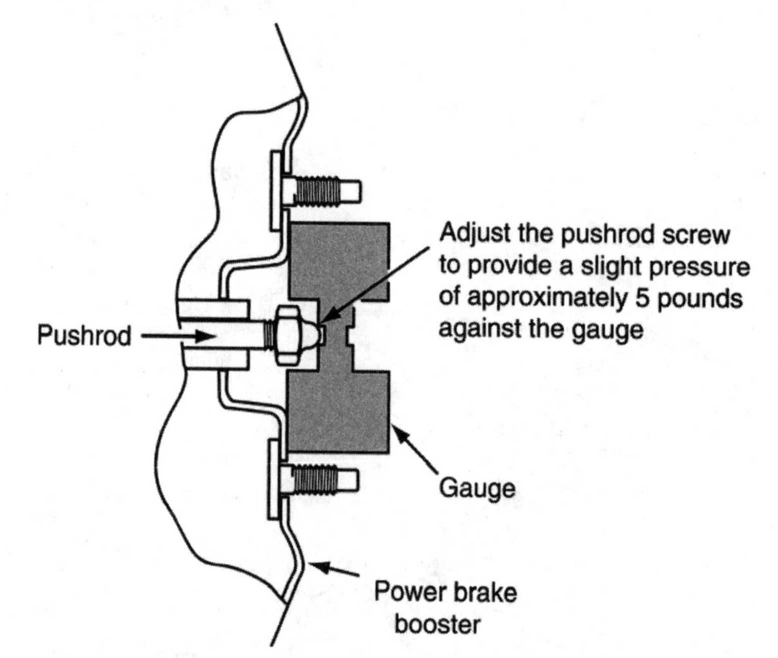

Task A.1.3 Measure and adjust master cylinder pushrod length.

Brake pedal height and pushrod adjustment are often overlooked when diagnosing a brake problem. Correct pushrod adjustment is very important for proper braking. The pushrod must have a very slight gap between the tip of the pushrod and its mating surface in the master cylinder. Most manufacturers' pushrod length can be adjusted by unlocking and rotating an adjustment nut that is on the tip of the pushrod. A pushrod that is too long may cause the master-cylinder piston to over-travel. This may damage the piston's seals and lead to a master-cylinder failure. A pushrod that is too long may also cause the piston seals to block off the compensating port, which may cause brake drag or lock-up. If a vehicle has had repeated master-cylinder failures, verify that the pushrod is not over-extended. A pushrod that is too short may limit the amount of brake pressure developed and reduce the overall braking ability of the vehicle.

Always check the brake light switch after making any adjustments.

Task A.1.4 Check master cylinder for failures by depressing brake pedal; determine needed repairs.

Three common complaints that can be caused by a faulty master cylinder are brake fade, spongy pedal, and excessive pedal effort. Before condemning the master cylinder, a complete brake inspection of all components must be performed.

If the brake pedal slowly sinks to the floor when applied and the vehicle is stopped with the engine running, there may be a leak in master cylinder cups. This is known as bypassing. Air leaking past the master cylinder piston cups will cause air to enter the fluid and cause a spongy pedal.

Swollen master cylinder cups may cause excessive pedal effort or can be the result of a corroded master cylinder bore. If the master cylinder reservoir cover seal is badly swollen, suspect brake fluid contamination from oil, transmission fluid, power steering fluid, etc. If fluid contamination is discovered, you must perform a complete inspection of the brake system. Contaminated fluid will damage all rubber components and seals in the brake calipers, wheel cylinders, and other brake parts.

You must flush all the contaminated fluid from the brake system and replace all seals and cups in the master cylinder, calipers, and wheel cylinders. In addition, any time the lines are disconnected from the master cylinder, or any other component, bleed the brake system to remove trapped air from the lines and master cylinder.

Task A.1.5 Diagnose the cause of master cylinder external fluid leakage.

External leaks from the master cylinder are usually easy to diagnose. Thoroughly clean the master cylinder first and check the fluid level. Check the condition of the reservoir seal and make sure the cover snaps tightly into place. If fluid level is low, it may be normal due to brake pad wear or it may be an indication of a leak.

Perform a complete visual inspection of the entire brake system to accurately determine any problems. If there is excessive fluid at the rear of the master cylinder, or if the paint on the booster is blistered, check for leaking brake fluid at the rear of the master cylinder. If a leak is present where the booster bolts up to the master cylinder, there will probably be brake fluid in the booster. The booster must be inspected and cleaned thoroughly and will have to be replaced.

Some master cylinders have a brake fluid reservoir, which is fitted into the main body of the cylinder. The seals of the reservoir and the cap must be inspected for possible leaks.

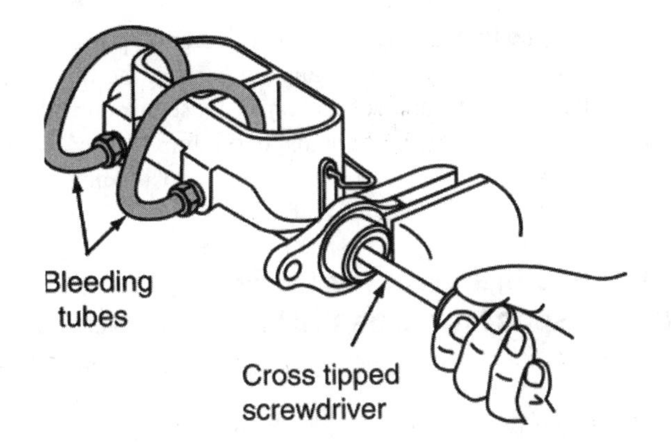

Task A.1.6 Remove and replace master cylinder; bench bleed and test operation and install master cylinder; verify master cylinder function.

Replacing the master cylinder is usually straightforward. Start by cleaning the exterior of the master cylinder and the area around the master. Remove the brake lines and plug the ports on the master cylinder and the brake lines. Disconnect any electrical wiring to the master. Remove the attaching nuts at the brake booster. If the vehicle has manual brakes, disconnect the pushrod from the brake pedal. Carefully lift the old master cylinder away from the vehicle to avoid spilling brake fluid on the vehicle's exterior. Use fender covers to avoid brake fluid damage to the paint. If brake fluid comes in contact with paint, wash immediately with water.

Before the replacement master is installed, it must be bench bled. Firmly secure the master cylinder in a vise and fill the reservoir with clean brake fluid. Make sure the master cylinder is level. Usually the new or rebuilt master will come with a pair of plastic fittings and hoses. Screw the fittings into the master cylinder ports. Attach one end of the hose to each port and place the other end of the hose into the reservoir. Make sure both hoses remain submerged in the brake fluid of the reservoir. Using a drift or dowel, slowly press the master cylinder piston in and out until the air is dispelled from the cylinder.

Step bore master cylinders require 75 to 100 psi to open the quick take-up valve. An option is to use a bleeder syringe to suck the air from the rear step bore of the master.

Mount the master cylinder back into the vehicle. Check pedal height and brake pedal free play. The entire brake system must be bled after the master is installed to ensure all the air is purged from the lines. Road test the vehicle and verify proper master cylinder and brake operation.

Task A.2 Lines and Hoses (3 Questions)

Task A.2.1 Diagnose poor stopping, pulling, or dragging caused by problems in the lines and hoses; determine needed repairs.

The master cylinder is connected to the wheel cylinders and calipers by a network of brake lines and hoses. Brake hoses create a flexible link between the steel line at the frame or body and the brake calipers or wheel cylinders.

The pressure and flow throughout the entire brake system must be within specification. Even a partial blockage in the lines or hoses can restrict the flow of fluid and cause problems. The applying brake pressure may be enough to force fluid through a restriction, but the affected brake may not release or will release slowly. This can cause brake drag. When the restriction affects one wheel, this can cause a pull to one side during braking.

Task A.2.2 Inspect brake lines and fittings for leaks, dents, kinks, rust, cracks, or wear; inspect for loose fittings and supports; determine needed repairs.

A comprehensive brake inspection of the brake lines and brake hoses should be part of any brake system diagnosis.

Kinks or dents in the brake lines may affect hydraulic pressure. Also inspect all fittings for leaks and secure connections. Check all attaching supports at the frame, undercarriage, rear suspension, and struts. Attaching brackets and supports may loosen or corrode over time. This may cause a shift in a brake line or hose and may cause damage to a line. A major concern for older vehicles, and in areas where salt is used during winter months, is corrosion and rust. Rusted brake lines can actually become porous and eventually leak brake fluid.

Task A.2.3 Inspect flexible brake hoses for leaks, kinks, cracks, bulging, wear, or corrosion; inspect for loose fittings and supports; determine needed repairs.

Because flexible brake hoses move with the natural movement of the steering and suspension, they must be inspected for wear. Front flexible brake hoses are especially susceptible to cracking and wear over time, as the wheels are turned left and right. Brake hoses should be inspected for bulging and wear marks caused by abrasion.

Check for the improper installation of brake hoses. Twisted brake hoses to calipers or loose brake hoses at the attaching supports are problems you may encounter. Whenever a brake hose is replaced, make sure a new sealing washer is used.

Task A.2.4 Replace brake lines hoses, fittings, and supports; fabricate brake lines using proper material and flaring procedures (double flare and ISO types).

The ends of brake lines are either double-flared design or have ISO (international standards organization)-type flared fittings. Copper tubing should never be used in a brake system. Copper is subject to fatigue and corrosion over time and can result in a hydraulic failure.

Factory replacement brake lines are often available. These lines are an exact fit with the proper end fittings. When brake lines are not available, it will be necessary to fabricate a new line. Replacement lines come in various lengths and diameter. Carefully choose the correct size brake line with the correct end fittings. Remove the old line and use it as template to fabricate the new line. A bending tool will be needed to bend tight angles. Although brake lines bend easily, be careful not to form any kinks in the process. Remember, always make sure of the end fittings: either double-flared or ISO-type flare. Also, watch for use of metric and/or fractional fittings on double-flare installations.

When it becomes necessary to flare the ends of the brake line, one of two different tools will be needed: the double-flaring tool and the ISO flare-forming tool. Never use compression fittings to join sections of brake line.

Always use a flare nut wrench to loosen or tighten brake line fittings. An open-end wrench can damage the end by *rounding* the corners of the fitting nut. Make sure all supports are intact. Loose brake lines or hoses may result in damage or failure.

Bleed the entire brake system after installing a new hose or line. Also, apply pressure to the brakes and inspect the repair for leaks. Road test the vehicle and confirm proper brake operation.

Task A.2.5 Inspect brake lines and hoses for proper routing and support.

All brake lines and hoses must be routed correctly. Inspect all lines and hoses and make certain that they are routed away from components that may interfere or rub, which can cause damage. Also, make sure brake lines and hoses are not routed near any components that generate excessive heat, such as an exhaust manifold, exhaust pipe, or catalytic converter.

Brake lines and hoses must be securely fastened to their supports. Unnecessary movement or vibration can lead to fatigue and eventual failure.

When replacing a flexible brake hose, make sure that it meets OEM safety standards and that the hose is the exact size as the original. A hose that is too long may rub on a chassis or steering component. If the hose is too short, it may stretch and break when the moveable component reaches the end of its travel. Always replace the sealing washer every time a brake hose is serviced.

Task A.3 Valves and Switches (3 Questions)

Task A.3.1 Diagnose poor stopping, pulling, or dragging caused by problems in the hydraulic system valve(s); determine needed repairs.

To provide evenly balanced braking and to warn of pending problems, many different valves are used in modern hydraulic brake systems. The most common valves are the metering valve, proportioning valve, and the pressure differential valve. These valves are sometimes housed in one assembly called the combination valve.

Faulty valves can cause brake drag, rear wheel lock-up, or front wheel lock-up upon braking. An in-depth review of these valves will follow in the upcoming paragraphs.

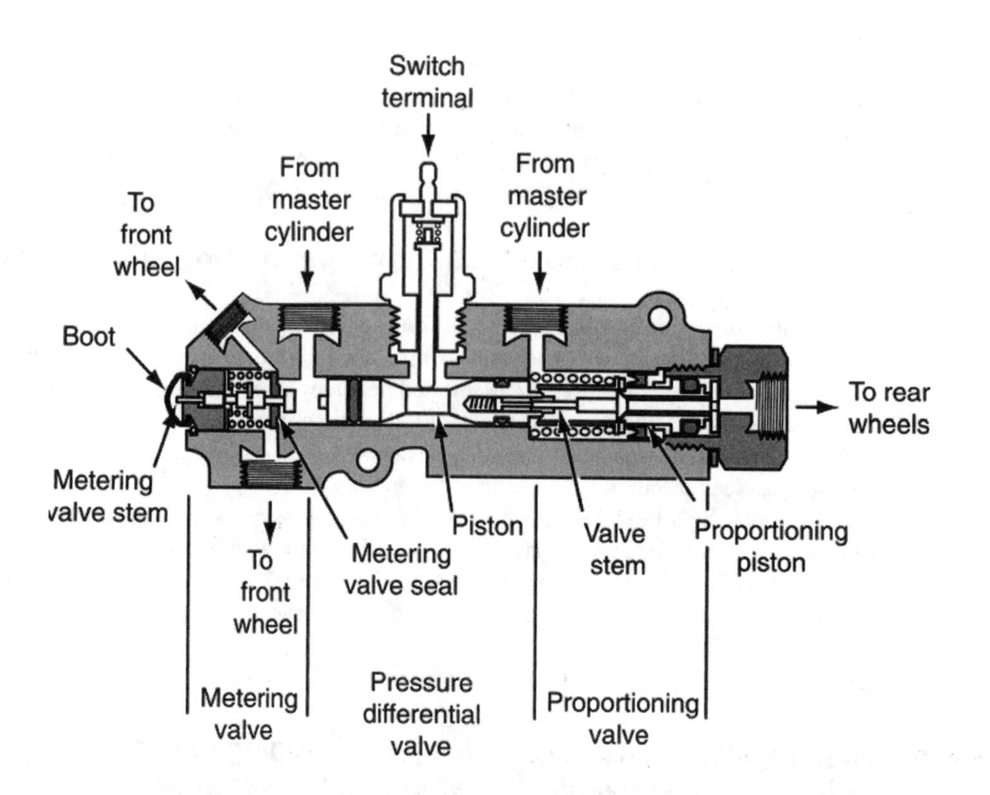

Task A.3.2 Inspect, test, and replace metering, proportioning, pressure differential, and combination valves.

The **metering valve** is used on vehicles with front disc and rear drum brake design. The metering valve delays the front disc brake operation until the rear drum brake shoes overcome the return spring tension. This will ensure simultaneous brake application of the front and rear brakes. Without this valve, the front disc brakes would apply too quickly.

A faulty metering valve can cause nose-diving and front wheel lock-up when the brakes are applied. During this condition, too much initial brake pressure is being applied to the front disc brakes. Premature front disc pad wear will also result from a faulty meeting valve. Inspect the valve for leaks.

The **proportioning valve** is also found on brake systems with front discs and rear drums. The proportioning valve controls and limits the pressure to the rear brakes.

Under mild braking, the pressure to all wheels is about the same. As pedal effort increases, the possibility of rear wheel lock-up exists. The proportioning valve prevents rear wheel lock-up under heavy braking. Vehicles with split diagonal brake systems utilize dual proportioning valves.

A malfunctioning proportioning valve may cause rear wheel lock-up under heavy braking, which may result in loss of vehicle control.

The **pressure differential valve** is used to warn the driver of a hydraulic pressure failure. The valve also incorporates a switch that will close and illuminate the dash warning brake light. A small piston floats inside the pressure differential valve cylinder, which separates the two halves of the brake system. The piston is normally centered inside the valve. When one side of the brake system develops a serious leak, the piston will be forced to the low-pressure side.

After the vehicle is repaired and brakes bled, the pressure differential valve should reset automatically. If the dash warning light does not go out, the valve may have to be re-centered. Check manufacturer's recommended procedure for centering the valve.

A **combination valve** incorporates the metering valve, proportioning valve, and the pressure differential valve all in one assembly. If any function of the combination valve fails, the entire assembly must be replaced.

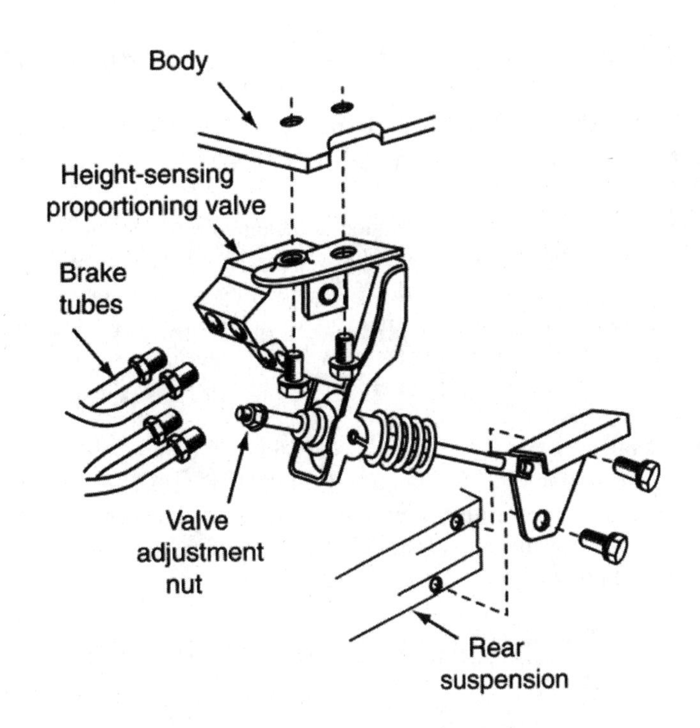

Task A.3.3 Inspect, test, replace, and adjust load or height sensing-type proportioning valve(s).

The load-sensing proportioning valve is mounted on the chassis, and a linkage is connected from the valve to the rear axle. When the load is light, the linkage positions the internal valve so brake pressure is reduced to the rear wheels during moderate brake applications. A heavy load on the rear suspension reduces rear chassis height. This action causes the linkage to move the valve in the height-sensing proportioning valve. Under this condition, the proportioning valve does not reduce pressure to the rear brakes during moderate brake application.

One type of load-sensing proportioning valve uses a bracket attached to the axle. A spring connects the bracket to a lever on the end of the proportioning valve, which is mounted to the vehicle chassis.

Weight transfer during hard braking will cause a change in chassis height. The increase in chassis height will decrease pressure to the rear wheels.

Any modification that alters the ride height will affect how the height sensing-type proportioning valve operates.

Task A.3.4 Inspect, test, and replace brake warning light, switch, sensor, and circuit.

When the pressure is equal in the primary and secondary sections of the master cylinder, the warning switch piston remains centered. In this position, the switch piston does not touch the switch pin. If the pressure is unequal between the primary and secondary master cylinder sections, the pressure difference moves the switch piston to one side. In this position, the switch piston pushes the spring-loaded switch pin upward and closes the warning switch light. This action illuminates the brake warning light.

The brake warning lamp circuit may be tested by grounding the warning switch with the ignition switch on. Under this condition, the bulb should light. If the bulb is not lit, check the fuse, bulb, and connecting wires.

Task A.4 Bleeding, Flushing, and Leak Testing (3 Questions)

Task A.4.1 Diagnose poor stopping, pulling, or dragging caused by problems in the brake fluid; determine needed repairs.

Brake fluid is the liquid that provides a means of transmitting hydraulic pressure from the master cylinder to the calipers and wheel cylinders. Brake fluid must meet federal requirements for viscosity, non-corrosive quality, water tolerance, high boiling point, and low freezing point.

DOT 3 fluid and DOT 4 fluid are hygroscopic, which means they absorb moisture. Over time the quality of brake fluid diminishes and may lead to internal corrosion of brake components. Corrosion can lead to sticking or seized calipers and wheel cylinders, which may cause brake drag. Corrosion can also cause a wheel cylinder piston, master cylinder piston, or caliper piston to seize in its bore.

Engine oil, transmission oil, or any other fluid other than brake fluid will contaminate the brake system and may swell rubber components of the brake system and cause brake drag. If a brake system has been contaminated with a foreign substance, it will be necessary to flush all the contamination from the brake system. Also, a careful inspection of the entire brake system is necessary because of the increased probability of damaged brake components.

Task A.4.2 Bleed and/or flush hydraulic system (manual, pressure, vacuum, or surge method).

Brake systems function on the principle of hydraulics, which states that a liquid cannot be compressed. However, air is compressible and if air is present in the brake system, the result will be a low and spongy pedal. Also, air contains moisture which may lead to component corrosion.

Air will enter the brake system when a hydraulic component is disconnected or if the level of master cylinder becomes too low. Bleeding the brakes is necessary to remove all the air in the brake system. Never reuse brake fluid.

There are four common methods for bleeding brakes: manual bleeding, pressure bleeding, vacuum bleeding, and surge bleeding. Whatever method is chosen, some basic procedures are the same. Check the level of the fluid in the master cylinder and maintain that level throughout the bleeding process. On all vehicles, follow the manufacturer's bleeding sequence. Bleeder screws are provided at the master cylinder, wheel cylinders, disc brake calipers, and sometimes on the combination valve. Using these bleeder screws, air is purged from the system by opening these screws and forcing fluid through them.

A pressure bleeder has an adapter connected to the top of the master cylinder reservoir. A hose is connected from this adapter to the pressure bleeder chamber in the top of the pressure bleeder. The pressure bleeder has an air chamber below the fluid chamber, and a diaphragm that separates the air and fluid chambers. Shop air is used to pressurize the air chamber up to 15 to 20 psi (103 to 138 kPa). If the brake system has a metering valve, this valve may have to be held open with a special tool. The bleeder hose is connected from each bleeder screw into a container partially filled with brake fluid. Open the bleeder screw until a clear stream of brake fluid is discharged.

During the vacuum bleeding procedure, the vacuum pump is connected to a sealed container. Another hose is connected from this container to the bleeder screw. A one-way check valve is connected in the hose from the bleeder screw to the container.

Operate the pump handle 10 to 15 times to create vacuum in the container. Open the bleeder screw until about 1 inch (25 mm) of brake fluid is pulled into the container. Repeat the procedure until the fluid coming into the container is free of air bubbles. During a manual brake bleeding procedure, connect a bleeder hose from a bleeder screw into a container partially filled with brake fluid. Keep the end of this hose submerged below the level of the brake fluid in the container. Each wheel caliper or cylinder must be bled in the vehicle manufacturer's specified sequence in any bleeding procedure. Wheel calipers may be tapped with a soft hammer to help remove air bubbles.

The surge method is a procedure that is sometimes used with manual brake bleeding. Surge bleeding is used in cases where it is difficult to remove trapped air from a wheel cylinder or brake caliper. In this method, the pedal is first pumped several times. The bleeder screw is then opened and, with a quick movement, an assistant depresses the pedal. The pedal is then released slowly. Wait a few seconds and repeat. On the last down stroke of the pedal, close the bleeder quickly.

Flushing the brake system involves continuing the bleeding process until all the fluid in the brake system is completely replaced. Flushing is recommended when brake fluid has been contaminated and is also recommended as a periodic brake service. Brake fluid absorbs moisture over time. Flushing removes the moisture and maintains the integrity of the brake components. If major brake components are replaced, flushing is highly recommended.

Task A.4.3 Pressure test brake hydraulic system.

Using pressure gauges, the brake system can be checked for correct hydraulic pressure at different components. Following manufacturer's specification, it is possible to determine if caliper, wheel cylinder, or part of the brake system is not receiving the correct pressure. Using pressure gauges can help in accurately diagnosing failed components.

The proportioning valves connected to the rear wheels may be tested by connecting pressure gauges on the master cylinder side and wheel cylinder side of each proportioning valve. When the brake pedal is applied with light pressure and the master cylinder pressure is below the split point pressure, both gauges should indicate the same pressure. If the brake pedal pressure is increased and the master cylinder pressure exceeds the split point pressure, the master cylinder pressure should exceed the wheel cylinder pressure by the manufacturer's specified amount.

After the brakes are bled, fill the master cylinder to the proper level. Following several firm applications of the pedal, check the calipers for leaks at the piston seals and the hose attachment areas.

Task A.4.4 Select, handle, store, and install proper brake fluids (including silicone fluids).

There are currently three types of brake fluid: DOT 3, DOT 4, DOT 5, and DOT 5.1. DOT 3, DOT 4, and DOT 5.1 are hygroscopic, which means they absorb moisture. All brake fluids must meet standards set by the Department of Transportation (DOT) and the Society of Automotive Engineers (SAE). The main difference of brake fluids is the point at which they begin to boil. DOT 3 is a glycol-based fluid that has a minimum boiling point of 401°F. DOT 4, which is also glycol-based, has a minimum boiling point of 446°F. DOT 5.1 is also a glycol-based fluid and has a 500°F boiling point. DOT 5 brake fluid is silicone-based, is not hygroscopic, and has a minimum boiling point of 500°F.

Due to their hygroscopic nature, DOT 3, DOT 4, and DOT 5.1 fluids have a limited shelf life. When a container of brake fluid is opened, the contents must be used as soon as possible. Always replace the cap on the container immediately. DOT 3 and DOT 4 will damage a painted surface, so take care when handling brake fluid.

The main advantage to DOT 5 silicon brake fluid is its high boiling point and the fact that it does not attract water. However, DOT 5 brake fluid is prone to aeration. Tiny air bubbles can form when the fluid is agitated. For this reason, DOT 5 fluid must never be used in a vehicle equipped with antilock brakes. DOT 5 fluid is also more difficult to bleed.

When adding brake fluid, always use fresh fluid from a new container and always check for the correct fluid for that vehicle. Most vehicles use either DOT 3, DOT 4, and DOT 5.1 fluid. These fluids can be interchanged, but really should not be mixed. DOT 5 fluid is not compatible with DOT 3 or DOT 4.

B. Drum Brake Diagnosis and Repair (5 Questions)

Task B.1 Diagnose poor stopping, pulling, or dragging caused by drum brake hydraulic problems; determine needed repairs.

A springy and spongy pedal is a condition where the brake pedal does not give firm resistance to foot pressure and feels elastic. It is normally caused by air in the hydraulic lines. Poor stopping ability may be caused by hydraulic problems such as contaminated fluid or air in the hydraulic system.

Brake drag may be caused by hydraulic problems such as contaminated brake fluid, inferior rubber cups in the master cylinder or wheel cylinders, or plugged compensating ports in the master cylinder. An obstructed brake line or hose may also be the cause of brake drag. A restricted brake hose or line may also cause a pull to one side or reduce stopping ability.

A seized wheel cylinder piston can cause brake drag if the piston is stuck in its bore. If the piston is stuck outward it can cause brake drag and premature brake shoe wear. A leaking wheel cylinder can cause air to enter the system and loss of fluid. This will result in a low and spongy pedal. Poor quality brake fluid or overheated brakes may lead to fluid boiling. At this point the fluid can be compressed and a temporary loss of proper braking will occur.

Task B.2 Diagnose poor stopping, noise, pulling, grabbing, dragging, or pedal pulsation caused by drum brake mechanical problems; determine needed repairs.

Brake squeal may be caused by bent backing plates, distorted drums, loose linings at shoe ends, improper lining position on the shoe, weak or broken hold-down springs, or loose wheel bearings. Brake chatter may be caused by improper brake adjustment; loose backing plates; contaminated linings; out-of-round, tapered, or barrel-shaped drums; cocked or distorted shoes; or loose wheel bearings.

Pedal pulsation may be caused by out-of-round brake drums.

Dragging or brake grabbing may be the result of contaminated brake shoes. Inspect the surface of the brake shoes for signs of brake fluid contamination or rear differential grease. Also, inspect wheel cylinders for seized pistons. If the wheel cylinder pistons do not retract when the brakes are released, this will result in brake drag. This drag will cause premature brake wear and overheated linings.

Task B.3 Remove, clean, inspect, and measure brake drums; follow manufacturer's recommendations in determining need to machine or replace.

Inspect brake drums for cracks, heat checks, out-of-round, bell mouths, scoring, and hard spots. Measure the inside drum diameter with a drum micrometer. If the drum diameter exceeds the maximum limit specified by the manufacturer, replace the drum. The drum diameter should be measured every 45 degrees around the drum. The maximum allowable out-of-round specified by some manufacturers is 0.0035 inch (0.09 mm). If the out-of-round exceeds specifications, the drum may be machined if the machining does not cause the drum to exceed the maximum allowable diameter.

Task B.4 Machine drums according to manufacturer's procedures and specifications.

When machining a brake drum, the drum must be installed securely and centered on the lathe. A dampening belt must be installed tightly around the outside of the drum to prevent the cutting tool from chattering on the drum. Many manufacturers recommend a rough cut tool of 0.005 to 0.010 inch (0.127 to 0.254 mm) and a finish cut tool depth of 0.005 inch (0.127 mm). After the machining procedure, the drum should be sanded to remove minor irregularities.

The surface of a freshly refinished drum contains millions of tiny particles. These particles can remain free on the surface or can become lodged in open pores of the cast-iron drum. If these particles are not removed, they can become imbedded in the linings. Once imbedded in the linings, the linings become a fine grinding stone, which will score the brake drum.

Start removing the particles by wet washing the lining with hot water, followed by wiping it with a lint-free cloth. Use compressed air to remove the remaining moisture.

Finally, use a lint-free cloth moistened with denatured alcohol or brake cleaning solution. Repeat the final step until the cloth shows no signs of dirt.

Task B.5 Using proper safety procedures, remove, clean, and inspect brake shoes/linings, springs, pins, self-adjusters, levers, clips, brake backing (support) plates, and other related brake hardware; determine needed repairs.

The parking brake strut is a bar between the brake shoes. When the parking lever is actuated, the parking brake strut pushes the leading brake shoe into the drum. Drum brakes contain specific components that are vital to the self-adjusting nature of the drum brakes. All brake return springs should be inspected for distortion and stretching. Brake shoes should be inspected for broken welds, cracks, wear, and distortion. If the wear pattern on the brake shoes is uneven, the shoes are distorted. Check all clips and levers for wear and bending. Inspect the brake linings for contamination with oil, grease, or brake fluid. Clean and lubricate adjusting and self-adjusting mechanisms.

A bent backing plate will not align the brake shoes properly. The misaligned shoes will result in brake drag or grab. Loose backing plate anchor bolts may cause brake chatter because the brake shoes can become misaligned if the backing plate moves.

While servicing a wheel cylinder, wash the parts in denatured alcohol or clean brake fluid. If the cylinder bore is pitted or deeply scored, replace the cylinder . The flat side of the piston should face the cups. Lubricate the piston cups with clean brake fluid before assembly.

If the brake shoes and linings have a slight blue coloring, this indicates overheating. In this case, the brake shoe adjusting screw springs and brake shoe hold-down springs should be replaced.

Linings must be replaced when contaminated with grease, oil, or brake fluid. Operating a vehicle for any length of time with worn brake shoes and linings will quickly result in scored brake drums. Always replace parts that are near the end of their service life. Never replace linings on one brake assembly without replacing those on the opposite wheel.

To check for leaks, pull back each wheel cylinder dust boot. Normally you will see a small amount of brake fluid present. This is not a cause for alarm as it acts as a lubricant for the piston. However, large amounts of fluid behind the boot indicate the fluid is leaking past the piston cups and an overhaul is in order.

Some wheel cylinders cannot be rebuilt. They are designed to be replaced whenever there is leakage or signs of damage.

Check the pads where the brake shoes rest. Look for any deep grooves in the brake shoe and lining contact pads that could resist brake shoe movement. Hand sand any grooves in the braking surface. If grooves are still present after sanding, the backing plate must be replaced. Any attempt to remove the grooves by grinding may result in improper brake shoe and lining-to-brake drum contact.

With manual brake adjusters, the brakes must be periodically adjusted to limit pedal movement and the increased need for brake fluid. However, the wear is so gradual that the driver may not realize how much out of adjustment the brakes have become.

Automatic brake adjusters generally take the worry out of keeping the brakes properly adjusted. The automatic adjusters adjust the clearance between the shoe linings and the drum.

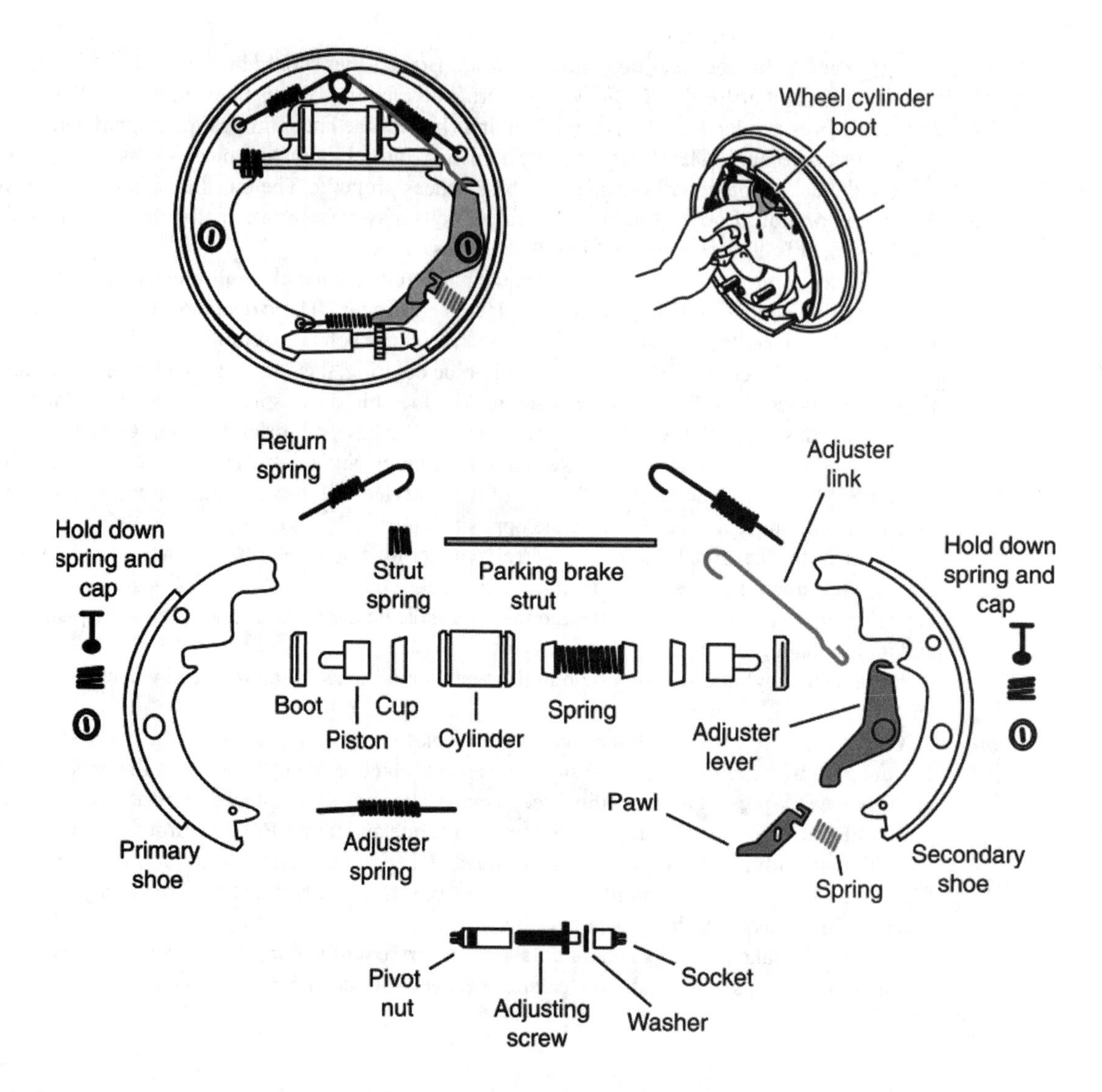

Task B.6 Lubricate brake shoe support pads on backing (support) plate, self-adjuster mechanisms, and other brake hardware,

Dry brake shoe ledges on the backing plate may cause a squeaking noise during brake applications. Lightly scored brake shoe ledges on the backing plate may be resurfaced and lubricated with high-temperature grease.

Whenever the rear brake shoe and linings are removed, check the parking brake rear cable and conduit tension.

Lubricate the ledges of the brake shoes that are positioned against the backing plates with the high-temperature grease recommended by the manufacturer.

Task B.7 Install brake shoes and related hardware.

When installing new brake shoes and hardware, the secondary shoe often has a longer lining than the primary and goes on the side of backing plate pointing toward the rear of the vehicle. The primary shoe faces toward the front of the vehicle. The self-adjuster cable is usually routed on the secondary shoe.

The adjuster must be installed in the proper direction so the starwheel is accessible through the backing plate opening or drum openings to allow for future removal of the shoes and to engage the self-adjuster lever. Be sure to lubricate the pads on the back plates and any pivot points with appropriate lubricant before installing the shoes and hardware.

Task B.8 Pre-adjust brake shoes and parking brake before installing brake drums or drum/hub assemblies and wheel bearings.

Before the brake drum is installed, position a brake adjusting gauge tool inside the brake drum and set the gauge for inside diameter. Next, position the gauge over the brake shoes and adjust the shoes to the gauge using the adjuster wheel. This procedure will pre-adjust the brake shoes and parking brake. This process is also simple and will save time.

On front drum applications and some larger trucks, it will be necessary to adjust the wheel bearing after mounting the drum in place. Follow manufacturer's specifications for correct wheel bearing torque and procedure.

Task B.9 Reinstall wheel, torque lug nuts; make final checks and adjustments.

One area that is a common source of comebacks is improper wheel installation. Wheel studs are designed to retain some elasticity when the wheel is bolted to the vehicle. If the wheels are over-torqued, this elasticity along with the integrity of the wheel mounting surface are no longer consistent which can result in brake noise, premature rotor warpage, or lug nut and stud damage. Recheck brake fluid level, road test, and verify proper brake function. Apply the parking brake and ensure proper parking brake function.

C. Disc Brake Diagnosis and Repair (10 Questions)

Task C.1 Diagnose poor stopping, pulling, or dragging caused by disc brake hydraulic problems; determine needed repairs.

Disc brake systems utilize a disc rotor, caliper assembly, and disc brake pads. The brake pads are positioned on both sides of the brake rotor and are squeezed against the disc rotor by hydraulic pressure acting on the caliper piston or pistons. It was once common to see vehicles built with disc brakes on the front wheels only; however, many cars and trucks today are equipped with four-wheel disc systems.

If the caliper piston is seized, proper brake action will be affected. On vehicles with front disc brakes, a seized caliper piston on one side can cause a pull if the brake pads are not being actuated properly. If the right front caliper is working properly, but the left front caliper piston is seized in its bore, the vehicle will pull to the right when the brakes are applied. If the caliper piston is seized and does not retract when the brake pedal is released, dragging of the brakes will occur. This may cause a pull to one side without applying the brakes and will cause uneven brake wear.

Another cause for dragging brakes could be faulty flex hoses. A restriction in the hose may allow pressure to apply the caliper properly, but not permit the pressure to release. This will keep the piston partially applied and cause the brakes to drag.

Always inspect the disc rotor for signs of bluing or glazing, which is an indication of overheating. Carefully inspect caliper hardware, pins, and slides. Also, closely examine the wear on the disc brake pads for each caliper. Each pad should wear evenly if pressure is being applied the same to each pad. Uneven pad wear may be an indication of a caliper with a hydraulic problem.

Task C.2 Diagnose poor stopping, noise, pulling, grabbing, dragging, pedal pulsation or pedal travel caused by disc brake mechanical problems; determine needed repairs.

There are two brake caliper designs: fixed and floating. Fixed caliper assemblies are rigidly bolted to the spindle or steering knuckle and do not move when the brakes are applied. Fixed caliper assemblies have pistons on both sides that apply hydraulic pressure against the brake rotor.

Floating (or sliding) calipers usually have only one piston that is actuated by hydraulic pressure. The caliper moves or slides as pressure is applied to the caliper piston. This action causes movement or sliding of the caliper and as a result will cause both pads in the caliper assembly to apply pressure to both sides of the brake rotor. These calipers will either float on pins or slide on a machined surface or support bracket. A common problem with floating calipers is tapered brake pad wear. Seized, sticking caliper pins, worn hardware, or corrosion on the support surface can cause the caliper not to float during braking. If one caliper is sticking or seized, it may result in a pull to one side as the brakes are applied.

Inspect all caliper hardware, which includes: anti-rattle clips, caliper pins, caliper slides, and shims. Worn or missing brake caliper hardware may cause noise, chattering, brake squeal, or brake pad rattling.

A pulsation felt through the brake pedal could be the cause of excessive disc rotor runout or loose wheel bearings. Thickness variation (parallelism) will also cause brake pulsation. Sticking or seized calipers can cause brake drag and may damage the rotors due to overheating. This situation may also overheat the brake pads and cause premature, uneven wear.

Task C.3 Retract integral parking brake caliper piston(s) according to manufacturer's recommendations.

Most vehicles with four-wheel disc brakes incorporate the parking brake mechanism within the rear brake caliper assembly. These calipers apply the piston hydraulically for normal braking and will apply the caliper piston mechanically when the parking brake is applied.

The caliper piston will turn and extend outward when the parking brake cable acts on a lever connected to the rear of the brake caliper. As the caliper piston turns outward in its bore, the pads clamp the rotor and prevent the vehicle from rolling.

To retract the caliper piston during rear pad replacement on this design caliper, you will need the aid of a special tool. Remove the caliper from its mounting position and remove the brake pads. Using the proper tool, rotate the caliper piston clockwise until it is seated back in its bore. The piston should rotate with some resistance but should retract under pressure applied from the tool. If the piston is seized, replace the caliper. Check parking brake function after reassembly. Adjust the parking brake cable as needed.

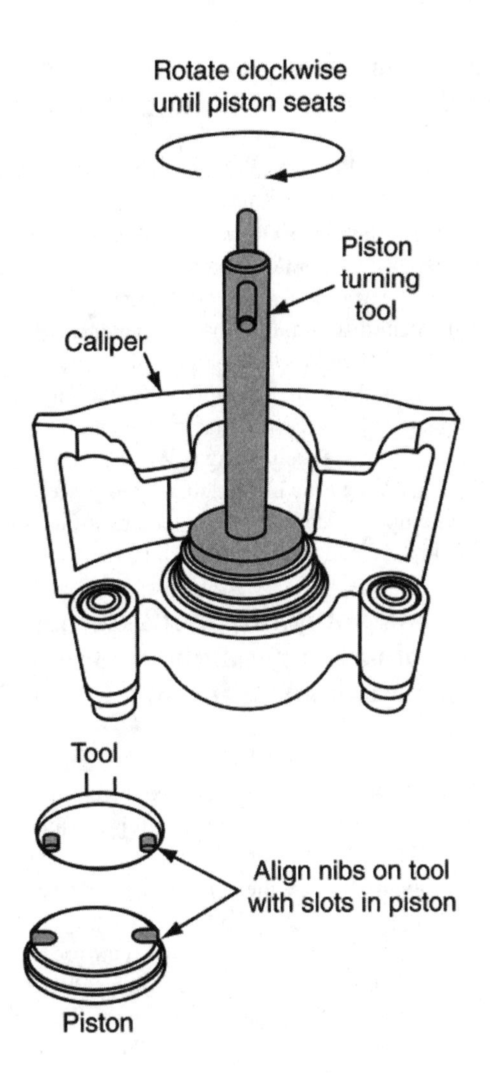

Task C.4 Remove caliper assembly from mountings; clean and inspect for leaks and damage to caliper housing.

After removing the caliper, carefully inspect all hardware, mounting contact points, caliper piston boot, and caliper housing. Thoroughly clean the caliper and mounting area. Inspect the dust boot for tears and for leaks from the piston. Also inspect the area where the brake hose connects to the caliper. Check the contact area where the caliper is positioned in the caliper anchor plate. Check for any physical damage to caliper housing or caliper. Any leaks from a caliper are unsafe and the caliper must be replaced.

Task C.5 Clean, inspect, and measure caliper mountings and slides/pins for wear and damage.

Inspect the caliper slide pins and sleeve assemblies for corrosion, and inspect all bushings for cuts and nicks. If damage to any of these parts is found, install new parts when the caliper is reinstalled. Do not attempt to polish away corrosion.

The caliper must move freely on the slide or pins with no binding. Inspect the surface area where the caliper is positioned on the anchor plate. All hardware, springs, anti-rattle clips, and pins must be cleaned and inspected. Worn or corroded parts must be replaced.

How the caliper mounts to the anchor or adapter is crucial for proper brake action. A loose caliper fit due to wear may cause brake noise or excessive caliper movement when the brakes are applied. The caliper may rock to one side rather than slide. This action is undesirable and must be corrected. Replacement of the caliper, slides, anchor plate, adapter, or caliper pins may be necessary if excessive wear is determined. All contact surfaces must be lubricated with appropriate brake caliper grease.

Task C.6 Remove, clean, and inspect pads and retaining hardware; determine needed repairs, adjustments, and replacements.

A scraping noise while braking may be caused by a pad wear sensor contacting the rotor. The scraping of the wear indicator indicates that the brake pads need replacement. If caught early, minimal damage will be done to the rotors. Remember that when resurfacing or replacing the rotors, the axle set must be either replaced together or have the same amount of machining done.

Shop manuals all specify minimum pad thickness, but the pads can only be measured if the unit is disassembled. Inspect disc brake rotors whenever the pads or calipers are serviced, or when the wheels are rotated or removed for any other work.

After removing the brake pads from the brake caliper, clean and inspect the brake pads, hardware, anti-rattle clips, and springs. Measure disc-pad lining and compare with manufacturer's specification. Check for tapered wear or cracked linings. Weak or broken springs or anti-rattle clips must be replaced. Most caliper slide pins have a coating to protect against corrosion. If the pins pitted or corroded, do not attempt to clean or polish. The pins must be replaced. Use appropriate brake grease on pins, slides, and brake pad contacting points.

Task C.7 Clean caliper assembly; inspect external parts for wear, rust, scoring, and damage; replace any damaged or worn parts; determine the need to repair or replace caliper assembly.

If a brake caliper is to be overhauled, it must be removed from the vehicle. Remove the brake hose and the attached caliper bolt. Carefully remove the caliper from its mounting. Discard the old copper washer used to seal the brake hose. A new sealing washer should always be used during caliper installation. Drain all brake fluid from the caliper and clean the exterior.

Before disassembly, make sure the caliper bleeder screws opens. If the bleeder screw is seized and cannot be freed, the caliper will have to be replaced.

The caliper piston can either be removed with air pressure or by using a piston removal tool. If using compressed air, clamp the caliper in a soft-faced vise and place a wooden block between the piston and the caliper body. Apply air pressure to the fluid inlet hole to force the piston from the caliper bore. **A word of caution:** the caliper piston will come out with extreme force; keep hands clear.

After the piston is removed, carefully pry away the dust boot and the caliper seal. Thoroughly clean the caliper bore and inspect for corrosion and pitting. Minor imperfections on the bore surface can be removed by honing. The caliper will need to be replaced if there is severe corrosion or damage. If the caliper bore is honed, check the piston-to-bore clearance. If the clearance is beyond manufacturer's specification, the caliper must be replaced. Inspect the groove where the piston seal is installed (seals are located either on the piston or in the caliper bore). If the piston groove is pitted or damaged, replace the piston. On calipers where the groove is located in the bore, carefully inspect for pitting and corrosion. Replace the caliper assembly if the seal grove or the caliper bore is damaged.

Carefully examine the caliper piston for damage and wear. On chrome-plated pistons, make sure the chrome is not peeling. Also inspect the piston for scoring or pitting. On phenolic pistons, check for cracks, chips, or gouges. Always replace a piston that is damaged. A new dust boot and piston seal will be used for reassembly of the caliper.

Before caliper reassembly, lubricate the caliper seal, piston, and bore with clean brake fluid. For calipers with seals located on the piston, install the seal and dust boot on the position. Then install the piston by pressing it into the caliper bore until it is fully seated. Use a suitable tool to seat the dust boot on the caliper. A C-clamp may be needed to help fully seat the piston. If the seal is located in the caliper bore, install the seal first. Then press the piston into the bore until it is fully seated. Using a suitable tool, install the dust boot.

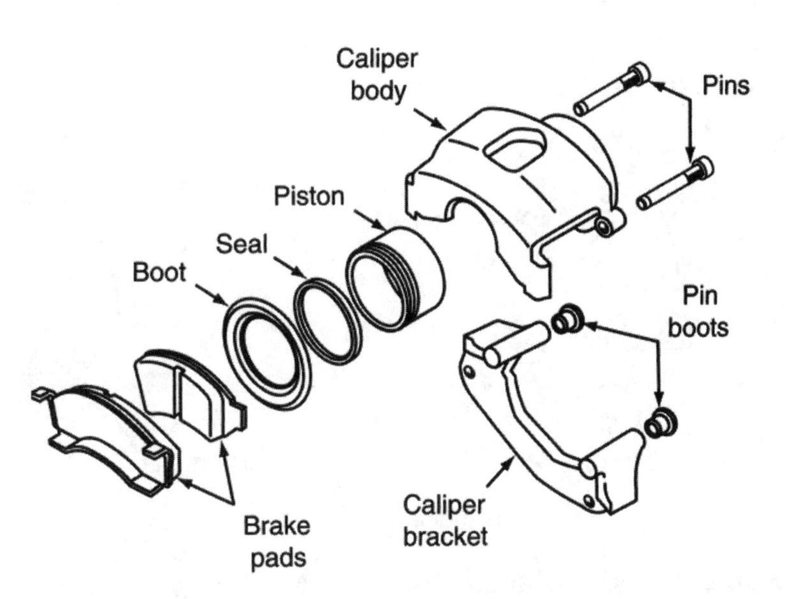

Task C.8 Clean, inspect, and measure rotor with a dial indicator and a micrometer; follow manufacturer's recommendations in determining need to index, machine, or replace the rotor.

The rotor is measured for parallelism or thickness variation; this measurement should be made at eight locations (every 45 degrees) around the rotor. These measurements should be taken near the center of the friction surface. Replace the rotor if the thickness variations exceed the manufacturer's specifications.

Remove the disc brake rotor; clean and inspect for damage. Check for pitting, grooves, heat cracks, or damage on the friction surface of the brake rotor. Blue spots on the rotor face indicate excessive heat. Replace the rotor if there is cracking.

Check all rotors for lateral runout (side-to-side wobble), parallelism (same thickness all the way around), and minimum thickness. Measurements can sometimes be made with the rotor still mounted on the vehicle with the caliper removed.

To check runout, mount a dial indicator to a solid surface. Place the indicator near the center of the friction surface and slowly rotate the rotor. See manufacturer's specifications.

Excessive parallelism can cause brake pulsation, brake shudder, and chatter. Also, if the disc rotor thickness is below minimum specification, discard the rotor. Minimum thickness specification is usually stamped into the rotor or can be found in service manuals.

If the disc rotor is not worn beyond specification, truing (or machining) can be performed. Machining the brake rotor involves the use of a brake lathe, which cuts both sides of the rotor in order to produce an even, smooth friction surface. Disc rotors must be machined in pairs to ensure smooth and even braking.

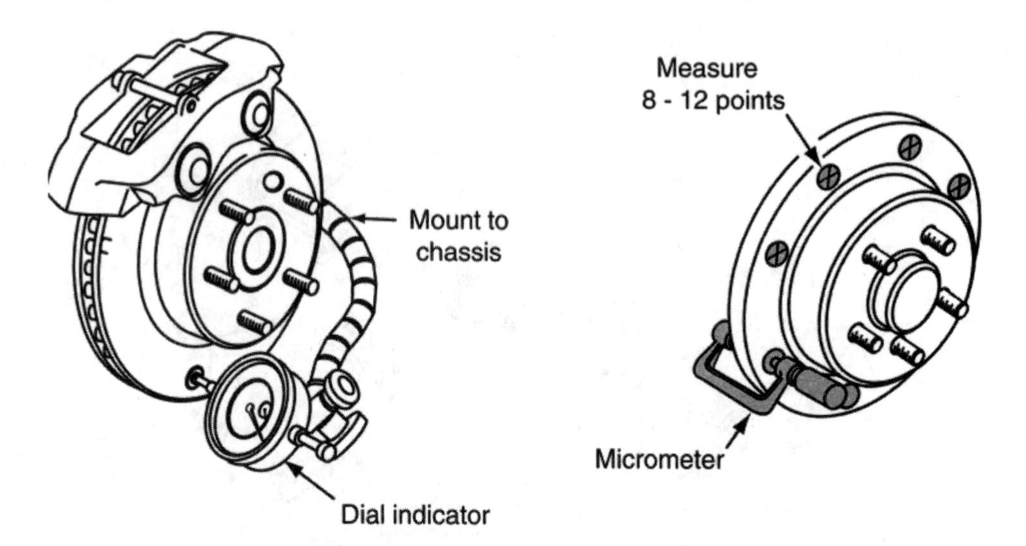

Task C.9 **Remove and replace rotor.**

Disc brake rotors either slide off the axle flange or are mounted in place by the wheel bearing. First, remove the caliper. Do not let the caliper hang by the flex hose. If the rotor is mounted to the axle flange, clean the flange before installing the new rotor. If the rotor is mounted to a spindle and secured with wheel bearings, clean all the grease from the spindle and bearings. Repack the wheel bearings and apply a small amount of wheel bearing grease to the spindle shaft. Following manufacturer's procedure, tighten and torque wheel bearings. Also, always replace the wheel bearing grease seal and use a new cotter pin. Check new rotors for runout, parallelism, and minimum thickness. Clean any oil film from the new rotor prior to installation.

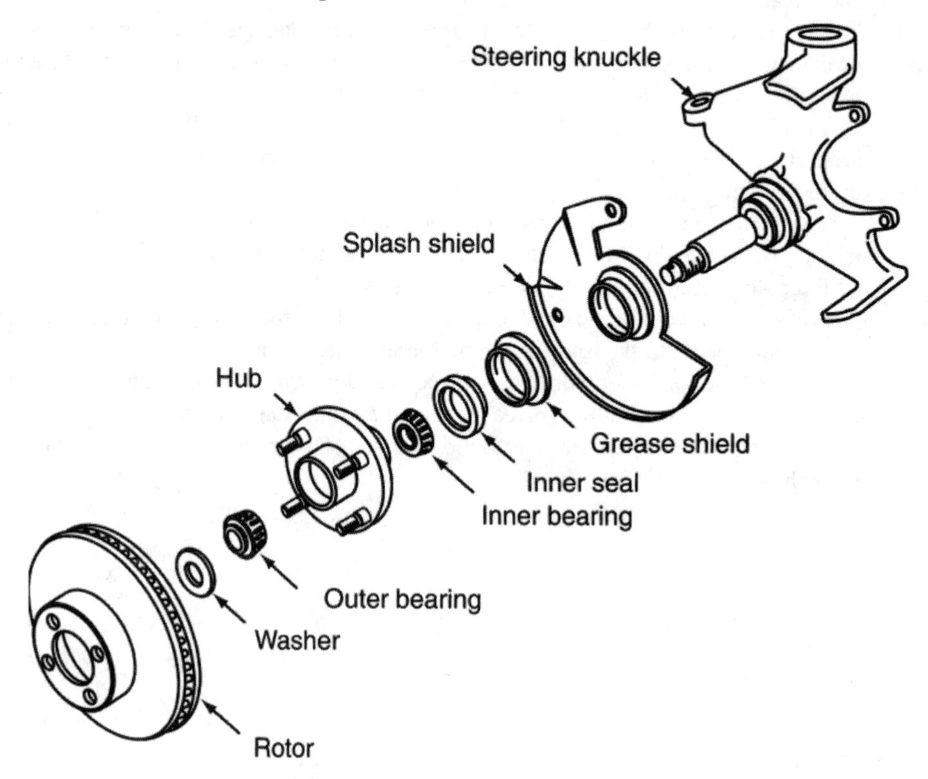

Task C.10 **Machine rotor, using on-car or off-car method, according to manufacturer's procedures and specifications.**

Machining the brake rotor involves the use of a brake lathe, which cuts both sides of the rotor in order to produce an even, smooth friction surface. Remove only enough metal stock to clean up the rotor surface. Never resurface one disc rotor. Disc rotors must be done in pairs to ensure smooth and even braking.

Brake rotors can either be machined on the car or off the car, depending on vehicle design and factory recommendations. An advantage to an on-car brake lathe is that it compensates for any variation in the axle flange, hub, or spindle. When machining rotors off the vehicle, a vibration damper must be positioned around the outside diameter of the rotor or on the rotor faces if the rotor is not vented. A fine grit (180) sanding pad should be used to sand the rotor surface after machining. This helps to knock the tops off of the sharp grooves left by the machining process. Equal amounts of metal must be removed from each side of the rotor.

To determine the approximate amount of metal to be removed, turn on the lathe and bring the cutting bit up against the rotating disc until signs of a slight scratch are visible. Turn off the lathe and reset the depth-of-cut dial indicator to zero. Find the deepest groove on the face of the rotor and move the cutting bit to that point without changing its depth-of-cut position. Now use the depth-of-cut dial to bottom out the tip of the cutter in the deepest groove. The reading on the dial now equals or is slightly less than the amount of material needed to be removed to eliminate all of the grooves in the rotor surface. For the best results with cuts that have a total depth greater than 0.015 inch (0.381 mm), take two or more shallow cuts rather than one very deep cut.

After machining the rotor, measure the rotor for minimum thickness. If the rotor was machined below minimum specification, discard the rotor. Also, give the rotor surface a non-directional finish using the proper lathe attachments. This will ensure proper brake pad-to-rotor surface contact. A sanding block using 180-grit sandpaper can also be used. Apply the sanding block for 30 seconds on each side of the rotor to achieve a smooth non-directional finish. Rotors must be washed thoroughly after machining. Soapy water works best. Simply submerge the rotor to remove leftover debris from machining. Brake cleaner and rags are not the preferred method because they usually leave behind metal debris that embeds in the new pads and can cause brake squeaks.

Task C.11 Install pads, calipers, and related attaching hardware; lubricate components following manufacturer's procedures and specifications; bleed system.

Clean off all caliper supports and sliding surfaces before installing the caliper assembly and pads. Lubricate sliding surfaces or pins with approved lubricant. It will be necessary to push the caliper piston back in order to install the brake pads. Open up the caliper bleeder screw first and push the caliper piston back using the appropriate tool. It is not recommended to push the caliper piston back with the bleeder screw closed. The fluid, along with any dirt, can be pushed back through the lines and back up to the master cylinder. Install the brake pads using new anti-rattle clips, shims, and other hardware to ensure the pads will sit properly and reduce the chance of squeal and noise. Tighten the caliper guide pins or bolts to the correct torque specification. Check the fluid level in the master cylinder and carefully pump the brake pedal until a firm pedal is felt. Bleed the system following the correct bleeding sequence and road test the vehicle to insure proper brake performance.

Task C.12 Adjust calipers with integrated parking brakes according to manufacturer's recommendations.

Some parking brakes on rear disc brake systems utilize a set of brakes shoes located inside the hub of the brake rotor. Adjusting the internal parking shoes is performed by rotating the starwheel adjuster with a tool through a slotted opening found in the face of the rotor or an opening found on the splash shield. Rotate the starwheel until correct shoe-to-drum clearance is obtained. Next, apply the parking brake pedal or lever, and remove the slack from the cables, if necessary. There should be no drag at the rear wheels when the parking pedal or lever is released. Apply and release the parking mechanism several times to ensure proper parking brake operation.

Rear disc brake systems that use the caliper piston to apply the parking function only require a cable adjustment.

Task C.13 Fill master cylinder to proper level with recommended fluid; inspect caliper for leaks.

Upon completing any brake service, refill the master cylinder with the correct, clean fluid. The brake fluid level on most master cylinders should be .25 inch (6.35mm) below the top. Many reservoirs are marked indicating full and refill levels. Check with the manufacturer for correct fluid type. Actuate the brake pedal to apply hydraulic pressure and inspect the calipers, lines, and hoses for leaks.

Task C.14 Reinstall wheel, torque lug nuts, and make final checks and adjustments.

An impact wrench should not be used to tighten wheel lug nuts since the excessive torque from this procedure may distort drums and cause excessive rotor runout.

All lug nuts must be tightened to correct torque specification. Over-tightening the lug nuts may damage the brake-rotor hub and cause excessive rotor runout. Before road testing, apply the brake and parking brake to check for correct brake function. Road test the vehicle and ensure proper brake performance.

D. Power Assist Units Diagnosis and Repair (4 Questions)

Task D.1 Test pedal free travel with and without engine running to check power booster operation.

Power-assist brakes contain a power booster connected to the master cylinder to reduce the required pedal travel and effort. With a relatively small amount of foot pressure, a great amount of hydraulic pressure can be produced. Most power brake boosters operate by engine vacuum and atmospheric pressure acting on a vacuum diaphragm. With the engine running, vacuum exists on both sides of the booster diaphragm. When the brake pedal is applied, a vacuum port is closed which allows atmospheric pressure to enter one side of diaphragm chamber. This action moves the diaphragm assembly and applies the master cylinder.

With the engine stopped, pump the brake pedal several times, and hold the pedal in the applied position. When the engine is started, the pedal should move slightly downward if the vacuum supply to the brake booster is normal. If the pedal does not move slightly downward, check the vacuum hose and the one-way check valve to the brake booster.

To check the vacuum brake booster for air tightness, operate the engine for two minutes and then shut off the engine. Pump the brake pedal several times with normal braking pressure. If the brake booster is operating normally, the pedal should go down normally on the brake application. The pedal should gradually become higher and harder to depress with each pedal application.

Task D.2 Check vacuum supply (manifold or auxiliary pump) to vacuum-type power booster.

Vacuum-assisted power brakes rely on a steady, adequate supply of manifold vacuum. Internal engine problems can affect power assist if manifold vacuum is lower than normal. Any vacuum leaks will also affect proper booster operation. A vacuum leak in the booster assembly can also occur. Low vacuum or a leaking booster will cause a hard brake pedal with little or no power assist. Engine vacuum should be 17-20 in. Hg at sea level at idle.

To check for proper vacuum, disconnect the booster hose from the manifold and connect a vacuum gauge. On vehicles with auxiliary pumps, test for vacuum at the pump with the engine running. Vacuum should be 15-22 in. Hg at sea level. Compare that reading right at the booster to make sure the booster is being supplied with the same vacuum.

Task D.3 Inspect the vacuum-type power booster unit for vacuum leaks and proper operation; inspect the check valve for proper operation; repair, adjust, or replace parts as necessary.

To check a one-way check valve in the brake booster vacuum hose, operate the engine at 2,000 rpm and then allow the engine to idle. Shut the engine off and wait 90 seconds. Pump the brake pedal five or six times. The first two pedal applications should be power assisted. If the first two brake applications are not power assisted, the one-way check valve is defective in the brake booster vacuum hose.

Insufficient manifold vacuum, leaking or collapsed vacuum lines, punctured diaphragms, or leaky piston seals can all result in weak power-unit operation. A steady hiss when the brake pedal is held down indicates a leak that can cause poor power-unit operation. Hard brake pedal is usually the first signal that the unit is on the way to complete failure.

Connect a vacuum gauge with a T-connection in the hose between the one-way check valve and the brake booster. With the engine idling, the vacuum should be 17 to 20 in. Hg at sea level (44 to 30.4 kPa absolute). If the vacuum is low, connect the vacuum gauge directly to the intake manifold and check this reading against the manufacturer's specifications.

Task D.4 Inspect and test hydroboost system and accumulator for leaks and proper operation; repair or replace parts as necessary; refill system.

The hydraulic power brake booster system (hydroboost) is used on some cars and trucks. Hydroboost systems use hydraulic pressure developed from the power steering pump to actuate the master cylinder. Some vehicles use belt-driven pumps, others use electric-driven pumps. A hydroboost system consists of a pump, booster assembly, master cylinder, and an accumulator.

As the brake pedal is depressed, the booster pushrod and piston are moved forward. This action causes the spool valve to move and allow fluid flow behind the power piston. As hydraulic pressure builds, it actuates the master cylinder.

The accumulator is used as a backup should the hydraulic power source fail. Accumulators are either spring-loaded or contain gas under pressure. The accumulator is filled with hydraulic fluid and is pressurized when the brakes are applied. If the engine stalls or a failure occurs in the power steering pump, the accumulator will have an adequate amount of pressurized fluid to provide one to three power-assisted brake applications. To test the accumulator, run the engine and then shut it off. Wait 5 to 10 minutes and test brake assist. If the accumulator is working correctly, you will have two to three assisted applications.

The inspection and testing of a hydroboost system must include a complete inspection of the power steering pump, belt, lines, hose connections, and the hydro-boost unit. Any leaks in the power steering system will affect hydroboost operation.

Aerated power steering fluid will also cause the hydroboost system not to function correctly. Low fluid level can cause a moan with a vibration in the pedal and steering column, usually experienced during parking or low speed maneuvers. It is important to remember that the pressure needed to actuate the hydroboost unit and master cylinder originates from the power steering pump assembly, which is usually operated by a belt, driven by the engine. A slipping belt or lower pump pressure will affect hydroboost operation.

Bleeding the power steering system will be necessary after repairs are made to the power steering system or to the hydroboost assembly. To bleed the system, perform the following procedure:

- Start the engine and apply the brake pedal several times while turning the steering wheel from stop-to-stop. Turn the engine off and apply the brake pedal several more times to deplete accumulator pressure. Check fluid level and add if needed. If the fluid is foamy, let the vehicle stand for a few minutes and recheck fluid level. It may be necessary to repeat the process until all air is purged from the system.
- With the wheels off the ground and the engine stopped, turn the steering wheel from stop-to-stop. Check fluid level and add if needed. Lower the vehicle.

E. Miscellaneous Systems (Pedal Linkage, Wheel Bearings, Parking Brakes, Electrical, etc.) Diagnosis and Repair (7 Questions)

Task E.1 Diagnose wheel bearing noises, wheel shimmy and vibration problems; determine needed repairs.

The wheel bearings perform a major role in effective brake function and steering performance. There are two different types of wheel bearing designs used on modern automobiles: the adjustable tapered roller bearing and the non-adjustable sealed roller or ball bearing.

Worn wheel bearings can cause a growling noise and vibration when the vehicle is driven. Checking for loose or worn wheel bearing should always be part of a complete brake inspection. A worn or improper wheel bearing adjustment can cause poor brake performance, poor steering, and rapid wheel bearing wear.

Tapered wheel bearings can be disassembled, cleaned, re-greased, and adjusted. Sealed roller or ball bearing wheel bearings cannot be serviced and are replaced as a unit.

Task E.2 Remove, clean, inspect, repack wheel bearings, or replace wheel bearings and races; replace seals; replace hub and bearing assemblies; adjust wheel/hub bearings according to manufacturer's specifications.

Tapered roller bearings are generally used on non-drive axles. The wheel bearings are mounted between a hub and a fixed spindle. To gain access to tapered wheel bearings, remove the wheel, brake rotor or drum, dust cap, cotter pin, spindle nut, and remove the outer wheel bearing. Remove the hub/rotor or hub/drum assembly. Pry the wheel bearing seal and remove the inner wheel bearing. Thoroughly clean the wheel bearing, hub assembly, spindle shaft, and races. Carefully inspect the wheel bearings, spindle, and races for signs of wear. Discard bearings showing any signs of wear, chipping, bluing, or galling. Repack the wheel bearings with high-temperature grease. Never repack a wheel bearing without first removing all the old grease. Insert the inner bearing into the hub and lightly lubricate the new wheel seal with grease. Tap the seal in place using a seal driver. Carefully install the hub/rotor or hub/drum assembly onto the spindle. Install the outer wheel bearing, washer, and spindle nut. Always reinstall a bearing in the same race.

If the wheel bearings need replacing it will be necessary to replace the bearing races also. Remove the bearing race from the hub using a brass drift or appropriate puller. Drive the new races into the hub until fully seated using a bearing race installer. If using a drift punch, tap the races a little at a time, moving the punch around the race to avoid cocking. Use a soft steel drift, never a hardened punch.

It is very important to properly adjust tapered wheel bearings. Always check with manufacturer's recommended procedure for the specific vehicle being serviced. There are two widely used methods for adjusting wheel bearings: the torque wrench method and the dial indicator method.

In the torque wrench method, rotate the wheel in the direction of tightening while the spindled nut is tightened to the specified torque. This initial torque setting seats the bearings in the bearing race. The nut is then loosened until it can be rotated by hand. The nut is then torqued, this time to lower specified value. Back the nut, if necessary, to install the cotter pin and lightly tap on the dust cap.

To adjust the wheel bearings using the dial indicator method, start by tightening the spindle nut while spinning the wheel, to fully seat the bearings. Loosen the spindle nut until it can be rotated by hand. Mount a dial indicator so the indicator point makes contact to the machined outside face of the hub. Firmly grasp the sides of the rotor or tire, and pull in and out. Adjust the spindle nut until the end play is within manufacturer's specification. Typical end play ranges from .001 inch to .005 inch. Install the cotter pin and dust cap.

Sealed ball and roller bearings are not serviceable and must be replaced when they are defective or have damaged grease seals. Removing this type of bearing involves pressing the bearing from the bearing hub out of the spindle or knuckle assembly. Carefully inspect the bearing, hub, and spindle assembly for wear. Press the new bearing into the spindle/knuckle assembly and torque the axle nut following the manufacturer's procedure.

Task E.3 Check parking brake system; inspect cables and parts for wear, rust, and corrosion; clean or replace parts as necessary; lubricate assembly.

The parking brake is a mechanical system integrated with the hydraulic brake system. It is intended to prevent the vehicle from rolling while it is parked. The rear wheels are held in place by a series of cables and linkages when either a parking brake pedal or hand lever is applied.

The parking brake cables must move freely when applied. Inspect the cables and all parking brake components for wear, binding, rust, and corrosion buildup. Check for cable wear at points of contact. A frayed cable must be replaced. Inspect the parking brake or lever assembly for binding and malfunction. Make sure the parking brake ratchets into place and locks into position. Check all the attaching cable hardware inside the brake drum. For disc brakes, inspect the hardware and lever at the rear of the caliper for wear and malfunction. Lubricate the cables and pivot areas as needed. Some cables are coated and cannot be lubricated.

Task E.4 Adjust parking brake assembly; check operation.

To check operation of the parking brake, raise the vehicle off the ground and apply the parking brake. The rear wheels should be locked in position. If the wheels can be rotated in the forward direction by hand with the parking brake applied, the parking brake may require adjustment. Before any adjustment is made, always check the condition of the brakes and drum brake adjustment. Worn rear brake shoes or brakes out of adjustment will affect the operation of the parking brake. If the parking brake is incorporated with the rear disc caliper, check the calipers for proper operation.

To adjust the parking brake assembly, raise the vehicle off the ground and apply the parking brake two or three clicks. There should be no slack in the cables and a slight drag felt at the rear wheels. If necessary, adjust the equalizer nut until all the slack is removed from the cables and a slight drag felt at the rear wheel. Release the parking brake. The parking brake should release and the wheels should turn freely. Next apply the parking brake about six clicks to see if the rear brakes lock. Make a final adjustment if needed. Check the service manual for specific procedure and number of required clicks to lock the rear wheel.

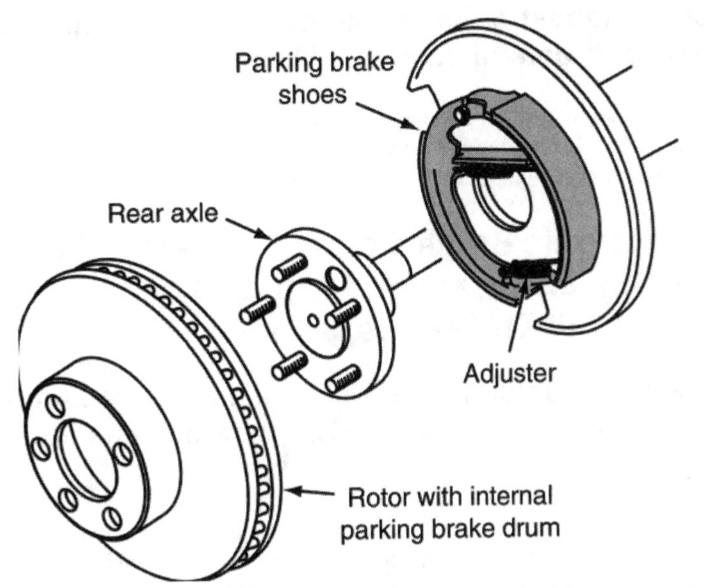

Task E.5 Test the parking brake indicator light, switch, and wiring.

Battery voltage is applied to a brake warning indicator light when the ignition is in the run mode, bulb test mode, or start mode. The parking brake switch supplies the ground for the warning light. The ignition switch supplies an additional ground for the brake warning light in the start or bulb test mode to provide a means to check proper bulb function.

A continually illuminated brake warning light may be caused by a grounded wire to the parking brake switch, a continually closed parking brake switch, or low fluid level in one section of the master cylinder reservoir. If an open circuit occurs in the wire to the parking brake switch, the brake warning light is not illuminated when the parking brake is applied.

When the parking brake pedal is applied, it closes a switch which completes an electric circuit to the brake indicator light in the instrument panel. The parking brake applied indicator light will then light when the ignition is turned on. The light goes out when the parking brake is released or the ignition is turned off. In some vehicles, this same indicator light may be used to alert the driver to problems in the antilock brake system.

Task E.6 Test, adjust, repair or replace brake stop light switch, lamps, and related circuits.

The brake light switch can be adjusted on some vehicles. The brake lights should operate when the brake pedal is depressed 0.25 inch (6.35mm). Battery power is supplied to one side of the switch, which is normally open. When the brake pedal is depressed, the switch closes, supplying battery power to the brake light bulbs.

If the brake lights do not operate, check for battery power to the brake light switch. If there is no battery power, check the fuse or check for an open wire to the brake light switch. If battery power is present at the switch, depress the brake pedal and check for battery power on the other side of the switch, which feeds battery power to the brake light bulbs. No power on the other side of the switch with the brake pedal depressed indicates a faulty switch.

If the switch is operational, and the brake lights do not operate, check for an open wire either on the battery feed side from the switch or ground side of the wiring harness at the brake light bulbs. Always consult the service manual for the specific wiring diagram when diagnosing the brake light circuit.

Some brake light switches contain multiple contacts to control cruise control function and torque converter clutch operation. Also, the brake light circuit on many vehicles is integrated with the signal switch or combination switch in the steering column. A defective signal switch or combination switch may affect the brake light circuit. On vehicles equipped with antilock brakes, the brake switch is also an input to the brake control module.

Task E.7 Inspect and test brake pedal linkage for binding, looseness, and adjustment; determine needed repairs.

Part of brake inspection includes checking the brake pedal and linkage for wear and proper function. Check for binding or worn bushings. Worn bushings or loose linkage may affect brake performance. If the vehicle was involved in an accident, check for a bent brake pedal.

F. Electronic Brake Control Systems: Antilock Brake System (ABS), and Traction Control Systems (TCS) Diagnosis and Repair (7 Questions)

Task F.1 Follow manufacturers' service and safety precautions when inspecting, testing and servicing ABS/TCS hydraulic, electrical, and mechanical components.

Many vehicles today are equipped with antilock brake systems (ABS). Although vehicle designs may differ, all ABS essentially function the same way. The ABS computer or control module monitors wheel speed, vehicle speed, and other vehicle functions. The wheel speed sensors send vital information to the ABS control module. When the control module senses that a wheel is about to lock up, it will regulate hydraulic pressure to the wheels in order to prevent lock-up. During a panic stop situation, the antilock brake system allows the driver to maintain directional control while providing maximum braking performance.

Most of the services done to the brakes of an antilock ABS are identical to those in a conventional brake system. There are, however, some important differences. One of these is the bleeding of the brake system. Always refer to the appropriate procedures in the service manual before attempting to service the brakes on an ABS-equipped vehicle. Before servicing an ABS, it is important that you understand the basics of electrical and electronic troubleshooting. Without this basic understanding, it will be difficult to follow the diagnostic procedures given in most service manuals.

Servicing antilock brakes requires following different safety precautions. Always refer to the service manual and review safety procedures for the specific vehicle being serviced.

Observe these common safety guidelines:

- Always wear safety glasses.
- Never open a bleeder valve or loosen a brake line on an integral ABS when the accumulator is pressurized.
- Remove the pressure from the accumulator by pumping the brake pedal 40 times with the ignition key in the off position.
- Never connect or disconnect any electrical connector or component with the key in the on position. This may damage the ABS controller.
- Never strike or tap on wheel speed sensor components. Striking these components may de-magnetize the wheel speed reluctor and will affect the signal strength of the sensor.
- Never mismatch tire sizes.
- DOT 5 fluid should never be used in ABS.

Task F.2 Diagnose poor stopping, wheel lock-up, pedal feel and travel, pedal pulsation, and noise concerns associated with the ABS/TCS; determine needed repairs.

The clicking noise during initial driving is a result of the ABS computer self-test mode in which the computer momentarily energizes the solenoid in the ABS system. On many systems, pedal pulsations are normal during ABS function; however, pedal pulsation during a normal stop when the ABS function is not operating may be caused by out-of-round drums or rotors with excessive runout.

Vehicles equipped with antilock brakes will have different braking characteristics during extreme braking than vehicles without ABS. During normal braking conditions, the antilock portion of the brake system does not function. However, wheel speed sensors continuously send information to the ABS control module. When wheel lock-up begins to take place, the control module will modulate hydraulic pressure through a series of solenoids to prevent wheel lock-up and maintain steering ability. Correct tire size plays a critical role in antilock operation. Different size tires can affect ABS operation and may cause wheel lock-up. During the ABS function process, the driver can misinterpret normal ABS operation as a brake problem. A pulsation felt through the brake pedal during ABS operation is normal. Also, a whine or ratcheting noise can be heard as the solenoids are modulating hydraulic pressure to the wheels. Some systems use an antilock pump assembly, which will initiate from time to time to maintain pressure in the accumulator. The driver may hear an audible whine from the pump. If the pump runs continuously or is activated for longer periods of time, this may indicate a faulty accumulator.

If no pulsation is felt under hard braking at speeds above 10 mph, suspect an antilock malfunction. Below 10 mph, the antilock system is generally disabled.

Task F.3 Observe ABS/TCS warning light(s) at start up and during road test; determine if further diagnosis is needed.

All ABS have some sort of self-test. This test is activated each time the ignition switch is turned on. You should begin all diagnostics with this simple test. To perform a typical ABS self-check sequence, place the ignition switch in the START position while observing both the red system light and the amber ABS indicator lights. Both lights should turn on. Start the vehicle. The red brake system light should turn off. With the ignition switch in the RUN position, the antilock brake control module performs a preliminary self-check on the ABS electrical system. During the self-check, the light remains on (three to six seconds) and then should turn off. If a malfunction is detected during the test, the light will remain on and the system will shut down.

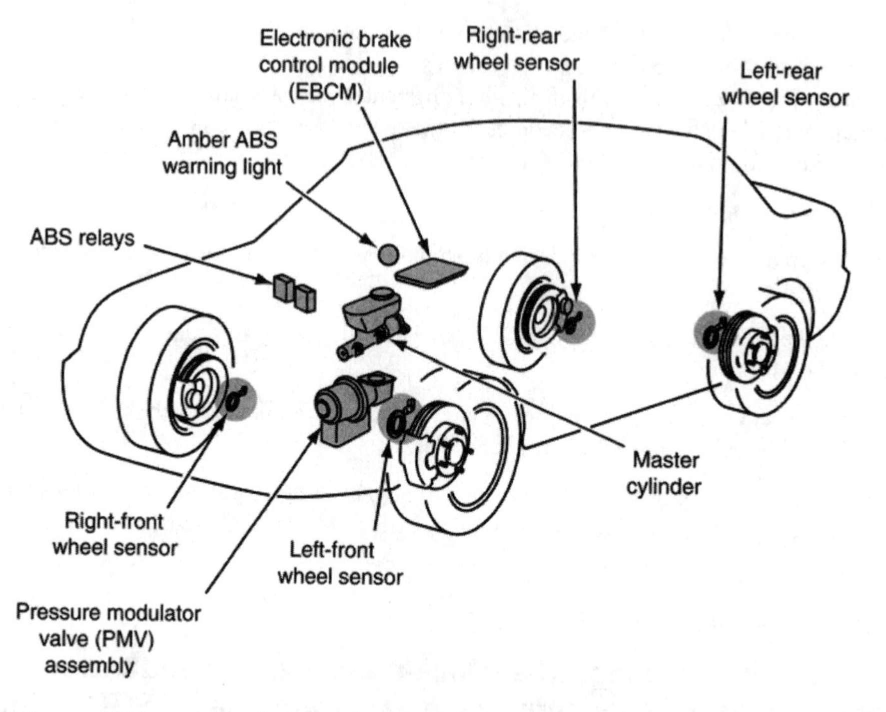

Task F.4 Diagnose ABS/TCS electronic control(s), components, and circuits using on-board diagnosis and/or recommended test equipment; determine needed repairs.

Most modern antilock systems have the ability to store an error message, called a fault code should a malfunction occur. By using a scan tool, the fault code can be retrieved from the computer memory. The fault code will direct you to troubleshooting procedures and aid in diagnosing the malfunction. On some systems, achieving the fault code is accomplished by counting the number of times the ABS warning light flashes. Always consult the appropriate service manual for specific diagnostic procedures.

Always begin your diagnosis with a complete visual inspection of the entire brake system and antilock system. Check fluid level, harness connectors, and wiring to components. If a fault code is stored in the ABS controller memory, follow the appropriate diagnostic procedure to diagnose the problem. Many systems allow the technician to view sensor and component data through a scan tool. Vital information, such as wheel speed sensor operation, can be viewed, compared, and analyzed. The use of a scan tool can greatly reduce the diagnostic time needed to solve tough ABS problems.

Task F.5 Bleed and/or flush the ABS/TCS hydraulic system following manufacturer's procedures.

Bleeding procedures for ABS will vary depending on the vehicle. Different ABS systems will require different bleeding methods. Some vehicles will require the use of a bi-directional scan tool to move solenoids or other components in a particular position. Many non-integral ABS can be bled using traditional methods. On most integral systems, a fully charged accumulator and special bleeding equipment is needed. Always refer to the service manual for specific recommended bleeding procedures.

Task F.6 Remove and install ABS/TCS components following manufacturer's procedures and specifications; observe proper placement of components and routing of wiring harness.

Some ABS utilize a pump, motor, hydraulic actuator, and accumulator with the master cylinder in one single unit (called integral ABS). Other systems connect the ABS hydraulics to the master cylinder and other components by means of high-pressure hoses. Most ABS components, such as wheel speed sensors, control modules, and hydraulic actuators are not serviceable and should be replaced if found defective.

When connecting and disconnecting any ABS components, make sure that the ignition key is in the off position. Some manufacturers advise removing the battery cables. Always make sure the ignition key is off and brake system depressurized before starting any repairs. Pumping the brake pedal at least 25 to 40 times will depressurize most ABS.

All wiring to any ABS component must be routed the exact same way it was removed. Do not alter the position, routing, or mounting of the harness.

Task F.7 Test, diagnose, and service ABS/TCS speed sensors (digital or analog), toothed ring (tone wheel), magnetic encoder, and circuits following manufacturer's recommended procedures (includes output signal, resistance, shorts to voltage/ground, and frequency data).

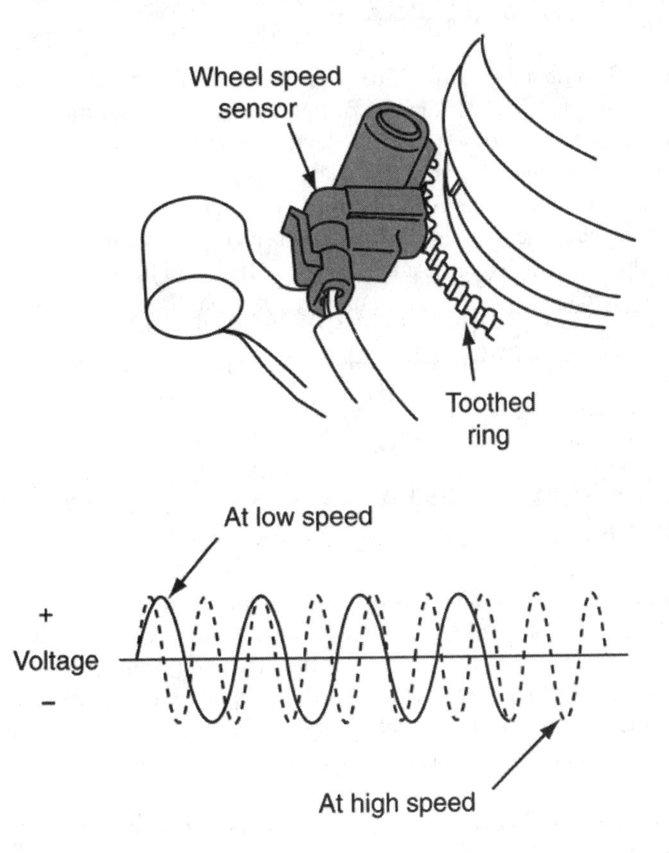

The wheel speed sensors continuously send signals to the ABS controller. The typical wheel speed sensor consists of a toothed ring (or tone wheel), made from ferrous metal, and a permanent magnet sensor. As the wheel turns, the wheel sensor produces a varying AC voltage based on the changing magnetic field. The strength of the signal is dependent upon the speed of wheel rotation. AC voltage will alternate to a maximum positive voltage to a maximum minimum voltage. This cycle is repeated many times and is known as a sine wave. By counting the number of cycles occurring each second the ABS controller can determine the frequency of the wheel speed sensor. The controller can use this information and compare the frequency of one wheel speed sensor against another. A wheel speed sensor decelerating faster than another indicates an impending wheel lock-up.

When testing a wheel speed sensor, start with a visual inspection of the sensor and wiring. Debris trapped around the sensor or ring can affect signal quality and strength. Also, if possible, check the air gap between the sensor and the toothed ring. A gap too wide may cause a weak signal or erratic readings. Inspect the toothed ring for debris, chips or cracks. Some sensors are mounted in the differential with the tone ring located on the ring gear. Metal particles can accumulate around the sensor from bearing and gear wear.

An ohmmeter can be used to check the resistance of the sensor. Compare the readings with the service manual. Probably, the most accurate way to test a wheel speed sensor is with a digital storage scope. A scope can test the quality of the signal and spot any intermittent glitches in the sensor signal.

The wheel speed sensor harness can also be a source of problems. The harness must be able to cleanly send information to the ABS controller. A misrouted sensor harness too close to a heat source or magnetic field can result in false information being sent to the controller. Corrosion in the harness or in a harness connecter can cause unwanted resistance in the wiring and a diminished wheel speed sensor signal. Always check the harness routing and check the continuity and quality of the harness wiring.

Task F.8 Diagnose ABS/TCS braking concerns caused by vehicle modifications (wheel/tire size, curb height, final drive ratio, etc.) and other vehicle mechanical and electrical/electronic modifications (communications, security, and radio, etc.).

All ABS computers are specifically programmed for the particular vehicle they are controlling. All aspects of the vehicle are taken into account: vehicle weight, transmission type, tire size, differential, suspension characteristics, electronic devices, and more.

Changing the tires and wheels to a size not recommended by the manufacturer can greatly affect ABS brake operation. The ABS controller is programmed to look for and expects to see specific information being sent by the wheel speed sensors. By changing the tire and wheel size, the signal will be altered and may cause a malfunction. It should be noted that many late-model trucks have the ability to change their programming for optional tire sizes. As with all things relating to vehicles every rule has an exception.

Any modification to the vehicle may affect ABS operation. If an electronic device, such as radio or an alarm, is installed, make sure the wiring harness does not interfere with the ABS harness. When replacing differentials and transmissions, always confirm the correct final drive ratio.

Task F.9 Repair wiring harness and connectors following manufacturer's procedures.

Always handle the ABS harness with care when testing, repairing, or servicing components. Never probe through the harness insulation and be very careful when backprobing a connector. Follow all guidelines prescribed in the service manual for testing components, connectors, and wiring. If a harness connector is faulty, do not attempt to repair. If a replacement connector is not available, replace the harness. If a section of harness wiring is damaged, it is recommended not to repair the wire, but to replace the damaged harness.

Sample Test for Practice

Sample Test

Please note the letter and number in parentheses following each question. They match the task in Section 4 that discusses the relevant subject matter. You may want to refer to the overview using the cross-referencing key to help with questions posing problems for you.

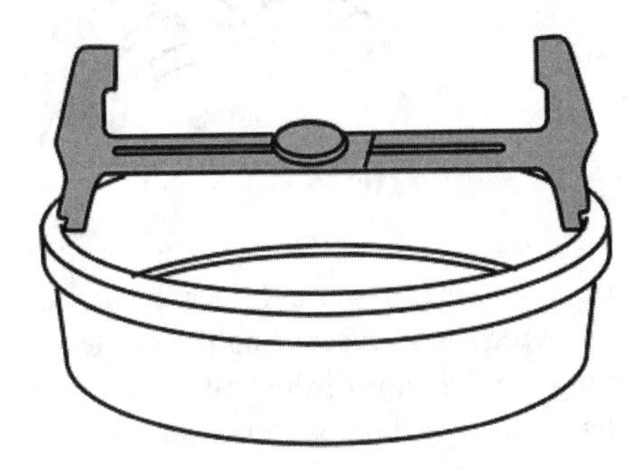

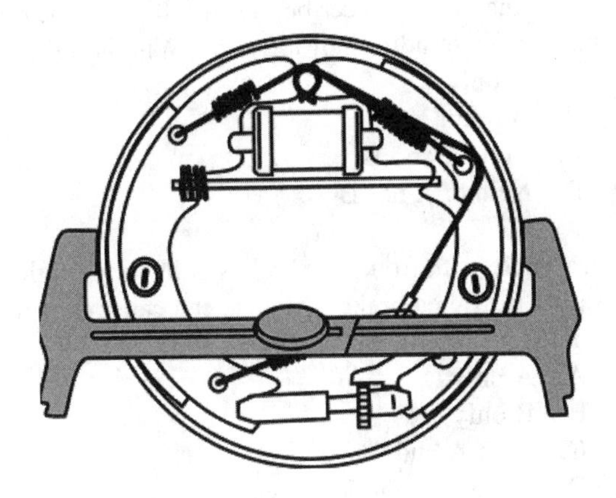

1. Technician A says the brake shoes shown in the figure are being adjusted to match the brake drum size. Technician B says the brake shoes are being measured for wear. Who is right?
   A. A only
   B. B only
   C. Both A and B
   D. Neither A nor B

(B.8)

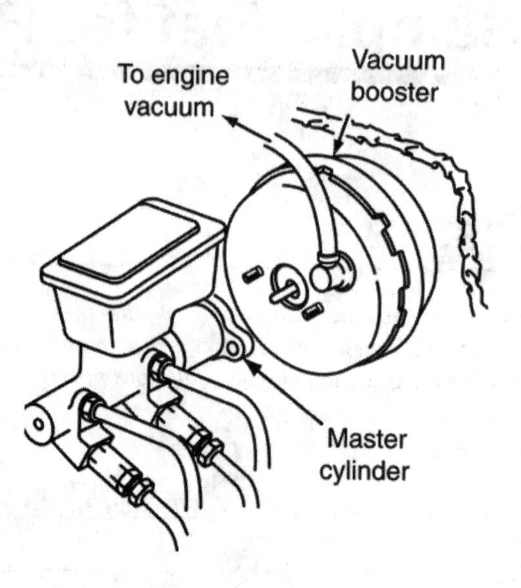

2. When removing the master cylinder, all of the following steps must be performed **EXCEPT:**
   A. removing the nuts securing the master cylinder to the booster.
   B. plugging the ends of the brake lines to prevent fluid loss.
   C. disconnecting the float level sensor.
   D. draining the master cylinder.

(A.1.6)

3. The front wheel bearings are being replaced on a rear-wheel drive vehicle. Technician A says to polish the bearing races before installing the new wheel bearings. Technician B says to carefully inspect the spindle shaft for wear. Who is right?
   A. A only
   B. B only
   C. Both A and B
   D. Neither A nor B

(E.2)

4. A vehicle with front disc brakes pulls to the right when the brakes are applied. Technician A says a restricted brake flex hose can cause this problem. Technician B says air in the brake lines can cause this problem. Who is right?
   A. A only
   B. B only
   C. Both A and B
   D. Neither A nor B

(A.2.1)

5. Technician A says that any time the brake lines are disconnected (or loosened) from the master cylinder, it has to be bled. Technician B says any time the brake system is worked on, the master cylinder must be bled. Who is right?
   A. A only
   B. B only
   C. Both A and B
   D. Neither A nor B

(A.1.4)

6. While discussing brake backing plates, Technician A says a bent backing plate may cause brake grabbing. Technician B says a loose anchor bolt may cause brake chatter. Who is right?
   A. A only
   B. B only
   C. Both A and B
   D. Neither A nor B (B.2)

7. All of the following statements are true about brake drum inside diameter measurements with a brake drum micrometer **EXCEPT:**
   A. the drum should be cleaned before measuring the diameter.
   B. if the drum diameter is less than specified, replace the drum.
   C. the diameter should be measured at multiple locations around the drum.
   D. inspect the drum for cracks, hard spots, and scoring. (B.3)

8. Technician A says that when a caliper is found to be malfunctioning due to Petroleum-contaminated brake fluid, the caliper must be replaced. Technician B says that if a brake component is damaged due to petroleum contamination, it may be necessary to replace all major components to insure proper brake function. Who is right?
   A. A only
   B. B only
   C. Both A and B
   D. Neither A nor B (C.7)

9. Which of the following should be used to clean the bores on wheel cylinders?
   A. Cleaning solvent
   B. Oil soap
   C. Hot water
   D. Denatured alcohol (B.5)

10. All of the following about brake hoses are true **EXCEPT:**
    A. some caliper components and hoses may be reverse threaded.
    B. most hoses are attached to the chassis and secured with a hose lock clip.
    C. cleaning the ends of the hoses prior to removal will reduce the possibility of dirt entering the system.
    D. replacement hoses can be longer than the original. (A.2.5)

11. A vehicle demonstrates brake drag. Technician A says fluid may be leaking past the master cylinder cups. Technician B says the stoplight and cruise control switch may require adjusting. Who is right?
    A. A only
    B. B only
    C. Both A and B
    D. Neither A nor B (A.1.1 and A.3.4)

12. During the teardown phase of a brake job, the technician discovers the front wheel bearings are badly damaged as a result of incorrect adjustment. Which brake pedal complaint would have lead to the teardown?
    A. Spongy
    B. Hard
    C. Rising
    D. Pulsating (E.1)

13. On an ABS with a high-pressure accumulator, the accumulator is depressurized by:
    A. pumping the brake pedal 25 to 40 times with the ignition switch off.
    B. pumping the brake pedal 10 times with the ignition switch on.
    C. loosening the bleeder screw on the accumulator with the ignition switch on.
    D. loosening the front wheel bleeder screws with the ignition switch on. (F.6)

14. When making final adjustments on the drum brakes, the star adjuster should be adjusted:
    A. until there is a slight drag on the wheel.
    B. until the shoes hold the wheel firmly in place and then backed off until there is a slight drag on the wheel.
    C. until there is a slight drag on the drum and then backed off until the wheel spins freely.
    D. one full turn to back the shoes further away from the drum. (B.9)

15. The rear wheels lock up too quickly during heavy braking on a vehicle with front disc/rear drum brakes. Which of these is the MOST-Likely cause?
    A. Air trapped in the master cylinder
    B. Faulty metering valve
    C. Defective brake booster
    D. Malfunctioning proportioning valve (A.3.1)

16. All four wheels spin freely on a rear-wheel drive vehicle that is raised off the ground. The brake pedal is depressed momentarily, then released. All wheels now spin freely except for the right front. The cause of the problem could be a:
    A. faulty proportioning valve.
    B. missing flex hose sealing washer.
    C. seized equalizer.
    D. restricted flex hose. (A.2.1 and A.2.2)

17. After installing a bench-bled master cylinder, Technician A says the entire system should be bled. Technician B says the system is ready for operation because the lines were capped. Who is right?
    A. A only
    B. B only
    C. Both A and B
    D. Neither A nor B (A.1.6)

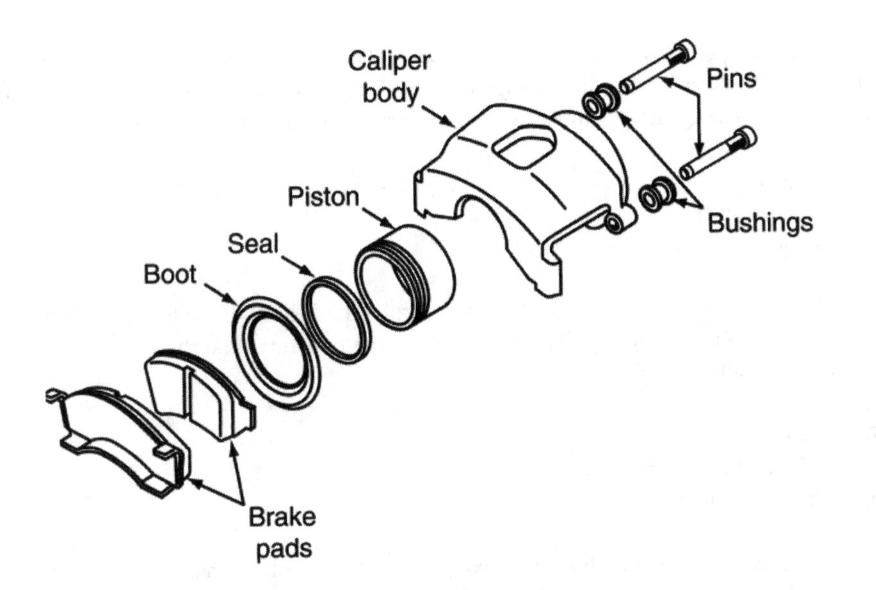

18. On a single piston floating caliper, the inside brake pad lining is worn out, but there is very little wear on the outside pad lining. The cause of the problem could be:
    A. worn caliper pins and bushings.
    B. a leaking caliper piston seal.
    C. a leaking brake hose.
    D. excessive rotor lateral runout. (C.7)

19. Technician A says a height-sensing-type proportioning valve will regulate pressure to the front and rear brakes with changes in ride height. Technician B says that during hard braking, the height-sensing-type proportioning valve will regulate pressure to the rear wheels. Who is right?
    A. A only
    B. B only
    C. Both A and B
    D. Neither A nor B (A.3.3)

20. When machining a brake rotor, Technician A says all brake rotors can be cut until the surface is smooth with no grooves. Technician B says after the first cut, rotate the rotor to reseat it on the brake lathe. Who is right?
    A. A only
    B. B only
    C. Both A and B
    D. Neither A nor B (C.10)

21. There are signs of paint removal and peeling on the power brake booster right below the master cylinder. Which of these would be the MOST-Likely cause?
    A. A leaking master cylinder piston cup
    B. A defective diaphragm in the brake booster
    C. Incorrect master cylinder pushrod adjustment
    D. Fuel being drawn into the booster by vacuum (A.1.5)

22. While discussing ABS service, Technician A says the high-pressure accumulator must be discharged before a brake line is disconnected. Technician B says some manufacturers recommend relieving the accumulator gas pressure before accumulator disposal. Who is right?
    A. A only
    B. B only
    C. Both A and B
    D. Neither A nor B

(F.1)

23. Technician A says that vacuum bleeding pulls brake fluid through the brake system. Technician B says that vacuum bleeding pulls brake fluid from a container and then empties it into the brake system. Who is right?
    A. A only
    B. B only
    C. Both A and B
    D. Neither A nor B

(A.4.2)

24. Technician A says that when the disc brake pads are replaced, the hardware and caliper slides should be cleaned and lubricated. Technician B says to replace all worn brake and caliper hardware. Who is right?
    A. A only
    B. B only
    C. Both A and B
    D. Neither A nor B

(C.5)

25. The antilock light comes on whenever the engine is started and stays on for a few seconds and then goes out. A road test reveals no unusual problems. Technician A says that there may be an intermittent antilock problem. Technician B says to retrieve trouble codes using the appropriate scanner. Who is right?
    A. A only
    B. B only
    C. Both A and B
    D. Neither A nor B

(F.3)

26. The left front brake does not lock on a vehicle with front disc/rear drum. The car also pulls to the right when the brakes are applied. A pressure gauge is used to check pressure at both front brake calipers. The pressure gauge reads close to 0 psi at the left front caliper. The right front caliper is within specification. The cause of this problem could be a:
    A. kinked brake line.
    B. seized left front caliper piston.
    C. seized right front caliper piston.
    D. binding pushrod.

(A.4.3 and C.1)

27. A fast-fill valve is a valve assembly that provides brake fluid to what area?
    A. The reservoir
    B. The spool area of the secondary piston
    C. The spool area of the primary piston
    D. The master cylinder outlet port

(A.1.2)

28. With the engine stopped, the technician pumps the brake pedal several times and then holds the brake pedal and starts the engine. The pedal moves slightly downward when the engine is started. This action indicates a:
    A. restriction in the power booster vacuum hose.
    B. defective check valve in the power booster vacuum hose.
    C. normal vacuum supply to the power booster.
    D. low intake manifold vacuum. (D.1)

29. After refinishing a brake drum, there are tiny metal particles remaining on the surface. These particles can be removed by:
    A. blowing the drum clean with compressed air.
    B. wiping the drum clean with a dry lint-free cloth.
    C. placing the hub side up and tapping with a dead blow hammer.
    D. wet washing with warm, soapy water and wiping it dry with a lint-free cloth. (B.4)

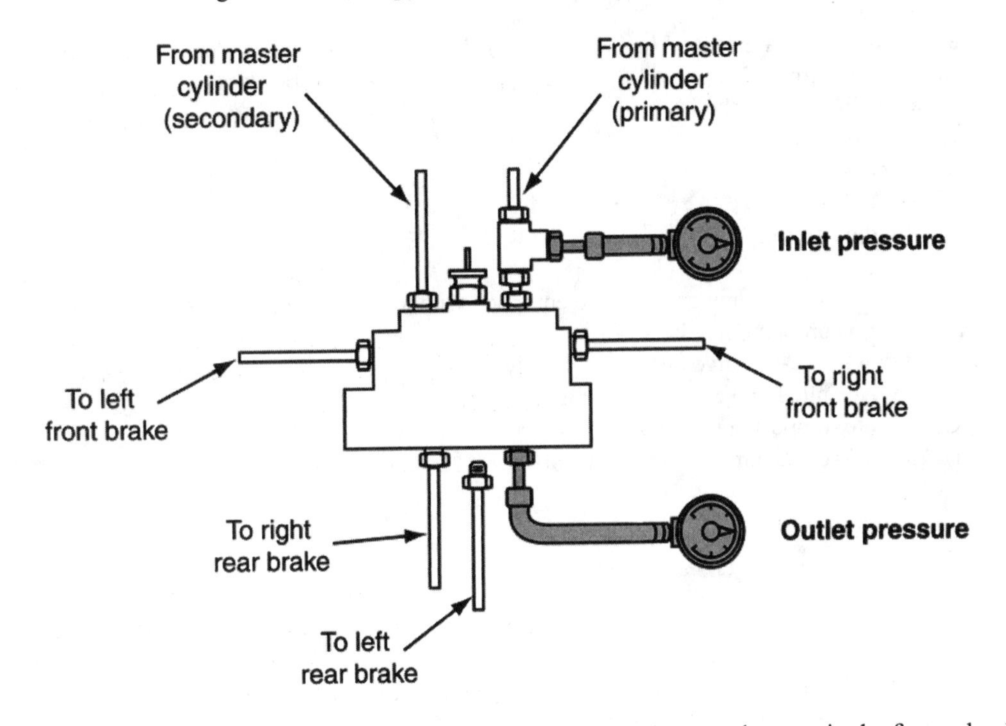

30. The pressure gauges connected to the proportioning valve input and output in the figure should indicate:
    A. higher pressure at the output compared to the inlet during a moderate brake application.
    B. the same pressure at the input and the output during a moderate brake application.
    C. lower pressure at the output compared to the input during a hard brake application.
    D. lower pressure at the output compared to the input during a light brake application. (A.3.2 and A.4.3)

31. On drum-type brakes, which of the following could cause a rising pedal on successive brake applications?
    A. Insufficient master cylinder pushrod clearance
    B. Master cylinder piston seal leak
    C. Reversed brake shoes
    D. Sticking wheel cylinder pistons (A.1.3)

32. A vehicle pulls to the left during a brake application. The cause of the problem could be the:
    A. left front brake linings are contaminated with grease.
    B. piston is seized in the right front brake caliper.
    C. master cylinder piston seals are swollen from contaminated fluid.
    D. secondary compensating port is plugged in the master cylinder. (C.1)

33. All of these statements about manual brake bleeding procedures are true **EXCEPT:**
    A. a hose is connected to the bleeder screw, and the opposite end of this hose is submerged in a container of brake fluid.
    B. apply the brake pedal with moderate pressure and then open the bleeder screw.
    C. when the bleeder screw is opened and the pedal goes to the floor, release the pedal and close the screw.
    D. repeat the bleeding procedure until the fluid escaping from the bleeder hose is free of air bubbles. (A.4.2)

34. Technician A says a damaged brake line may be repaired with a short piece of line and compression fittings. Technician B says the necessary brake line bends should be made with a tubing bending tool. Who is right?
    A. A only
    B. B only
    C. Both A and B
    D. Neither A nor B (A.2.4)

35. The front disc brakes drag on a vehicle equipped with a step bore master cylinder. A possible cause for this problem may be:
    A. a quick take-up valve opening too early.
    B. an over-filled brake fluid reservoir.
    C. a faulty proportioning valve.
    D. a blocked vacuum booster check valve. (A.1.2)

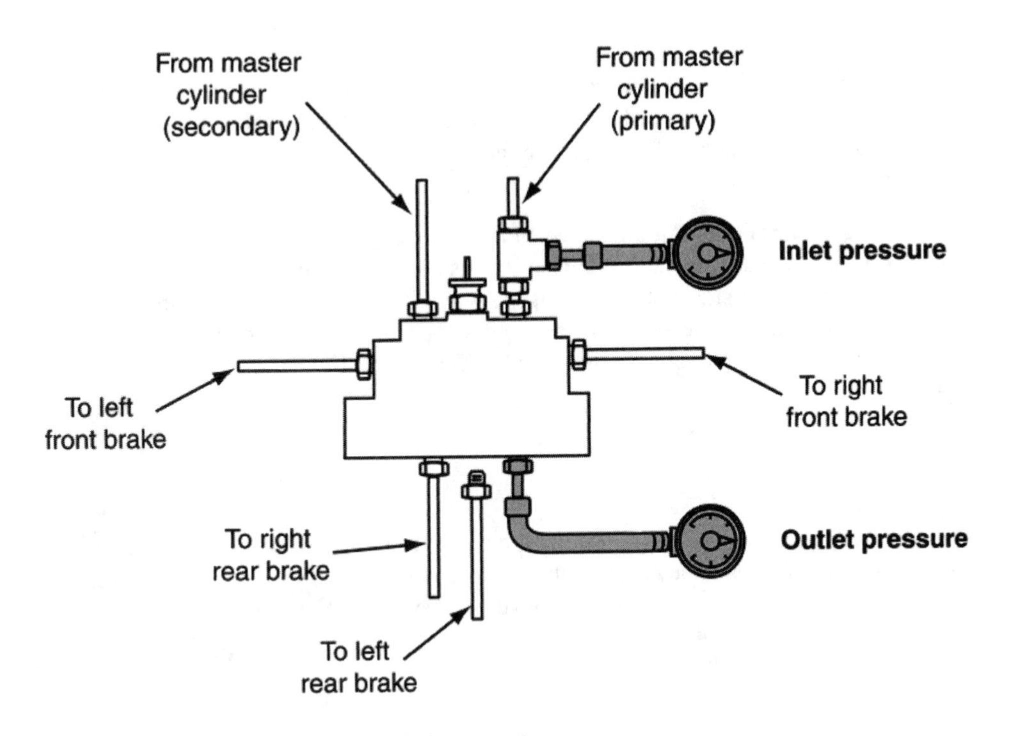

36. In the figure, which component is being tested:
    A. A wheel cylinder
    B. A master cylinder quick take-up valve
    C. A proportioning valve
    D. A caliper

(A.3.2 and A.4.3)

37. Evidence of brake fluid around the caliper boot area indicates:
    A. an over-filled master cylinder.
    B. a bad caliper seal.
    C. restricted brake lines.
    D. worn caliper pins and bushings.

(C.4)

38. While discussing brake hardware service, Technician A says dry shoe ledges on the backing plates may cause a squeaking noise during brake application. Technician B says slightly scored shoe ledges on the backing plates may be resurfaced and lubricated with high-temperature grease. Who is right?
    A. A only
    B. B only
    C. Both A and B
    D. Neither A nor B

(B.6)

39. A fault code relating to the right front wheel speed sensor circuit is retrieved using a scan tool. Technician A says that a wheel speed sensor can be tested using a digital storage scope. Technician B says replace the right front wheel speed sensor, clear the code, and road test the vehicle. Who is right?
    A. A only
    B. B only
    C. Both A and B
    D. Neither A nor B

(F.4)

40. To test the brake warning light circuit,:
    A. drain the master cylinder, start the engine, and watch for the warning light to illuminate.
    B. disconnect the switch, start the engine, and apply moderate brake pressure.
    C. replace the switch because it may be faulty.
    D. ground the warning switch wire with the ignition on. (A.3.4)

41. The stop light bulbs do not illuminate when the brake pedal is depressed. The stop light circuit fuse was checked and is good. Technician A says to depress the brake pedal and check for battery power on both sides of the stop light switch. Technician B says if battery power is present on both sides of the switch when the brake pedal is depressed, check for a short to ground in the harness to the brake lights bulbs. Who is right?
    A. A only
    B. B only
    C. Both A and B
    D. Neither A nor B (E.6)

42. A vehicle experiences poor stopping, and the pedal feels springy and spongy during brake application. Technician A says the vents in the master cylinder cover may be plugged. Technician B says there may be air in the hydraulic system. Who is right?
    A. A only
    B. B only
    C. Both A and B
    D. Neither A nor B (B.1)

43. A vehicle with ABS is brought in due to a pedal pulsation. Technician A says some pedal pulsation is normal during hard braking with ABS. Technician B says the brake rotors may have excessive runout. Who is right?
    A. A only
    B. B only
    C. Both A and B
    D. Neither A nor B (F.2)

44. A metering valve delays pressure to which set of brakes under which braking condition?
    A. The rear brakes under light braking
    B. The rear brakes under heavy braking
    C. The front brakes under light braking
    D. The front brakes under heavy braking (A.3.2)

45. Brake pedal pulsation is experienced while braking. Technician A says the outer constant-velocity joint may be worn. Technician B says the front wheel bearing may be worn and loose. Who is right?
    A. A only
    B. B only
    C. Both A and B
    D. Neither A nor B (C.2)

46. Technician A says the light comes on when there is a pressure difference on one side of the brake pressure differential switch. Technician B says the light comes on when pressure is the same on both sides of the brake pressure differential switch. Who is right?
    A. A only
    B. B only
    C. Both A and B
    D. Neither A nor B (A.3.2)

47. A diagnostic trouble code (DTC) representing the L/R wheel speed sensor is obtained in the ABS. This DTC may be caused by:
    A. an open circuit at the R/R wheel signal sensor.
    B. a grounded circuit on the L/R wheel sensor.
    C. a larger tire than specified by the manufacturer on the R/F wheel.
    D. an open circuit in the electronic brake control. (F.7)

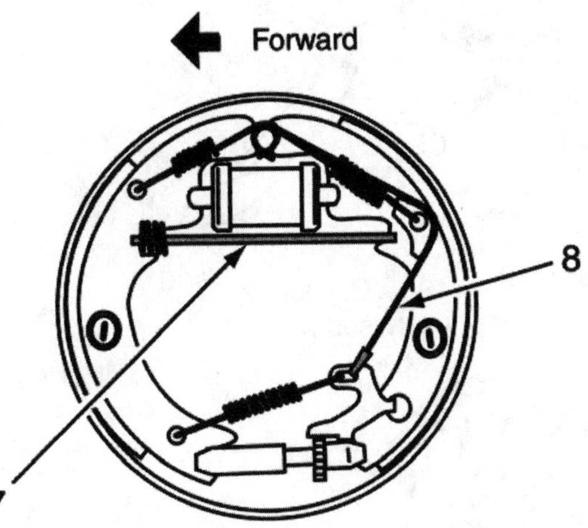

48. Refer to the illustration. Technician A says if component number 17 is omitted during brake hardware assembly, the brake will grab on the drum. Technician B says if item number 8 is omitted during brake assembly, there will be no self-adjusting action. Who is right?
    A. A only
    B. B only
    C. Both A and B
    D. Neither A nor B (B.5)

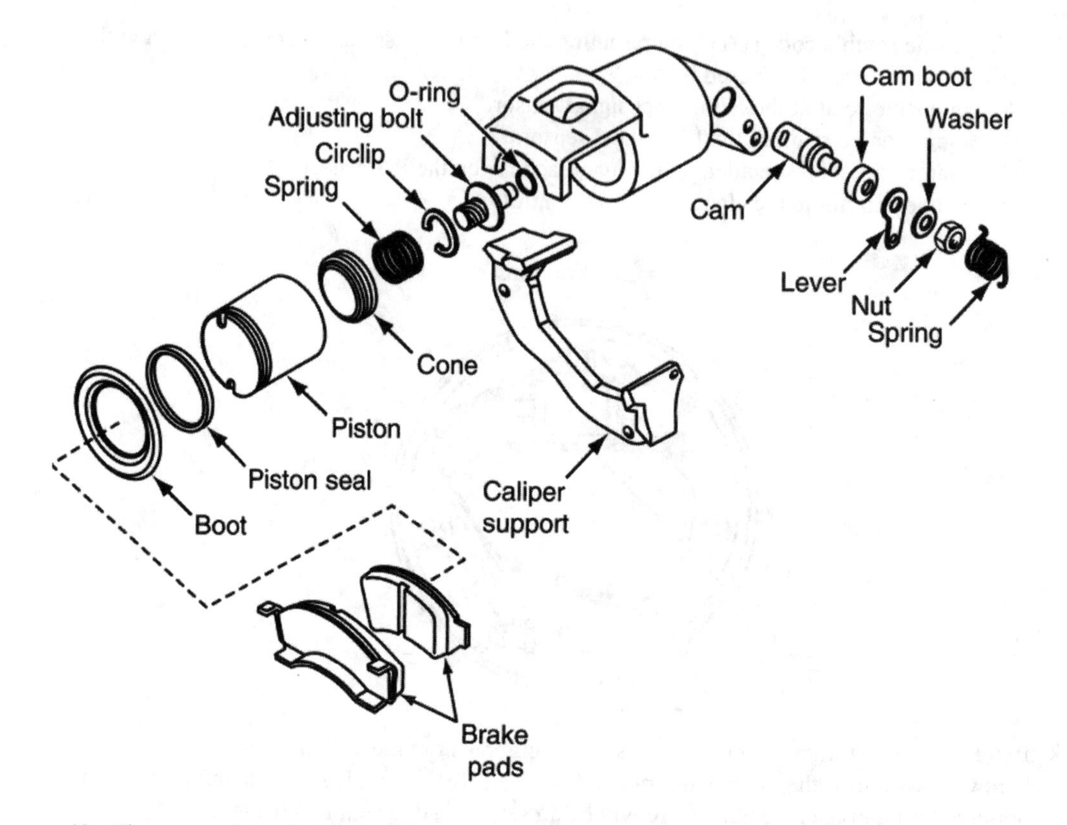

49. The rear pads are being installed on a vehicle with integral parking brakes. Technician A says a special tool is needed to retract the caliper piston. Technician B says parking brake function should be checked after the pads are installed. Who is right?
    A. A only
    B. B only
    C. Both A and B
    D. Neither A nor B (C.3)

50. Technician A says excessive lug nut torque may cause rotor runout. Technician B says an impact wrench may be used to tighten lug nuts. Who is right?
    A. A only
    B. B only
    C. Both A and B
    D. Neither A nor B (C.14)

51. A hissing sound is heard from under the dash when the brakes are applied with the engine running. Technician A says a faulty booster check valve can cause this. Technician B says a leaking vacuum booster is the cause. Who is right>
    A. A only
    B. B only
    C. Both A and B
    D. Neither A nor B (D.3)

52. The power steering operates normally on a hydroboost-equipped vehicle. During a hydroboost brake test, the brake pedal is pumped several times with the engine not running. When medium pressure is applied to the brake pedal and the engine is started, the brake pedal height remains unchanged. Technician A says the power steering pump pressure may be low. Technician B says the accumulator may be defective in the hydroboost unit. Who is right?
    A. A only
    B. B only
    C. Both A and B
    D. Neither A nor B (D.4)

53. A technician applies the parking brake and releases it. The rear brakes do not release and the wheels are locked with the parking brake pedal in the retracted position. The cause of the problem could be:
    A. binding parking brake cables.
    B. missing self-adjusting cable.
    C. seized wheel cylinders.
    D. defective ABS accumulator. (E.3)

54. Before adjusting the parking brakes, the technician should:
    A. bleed the front brakes.
    B. check the tire pressure.
    C. visually inspect the system.
    D. fill the master cylinder with clean brake fluid. (E.4)

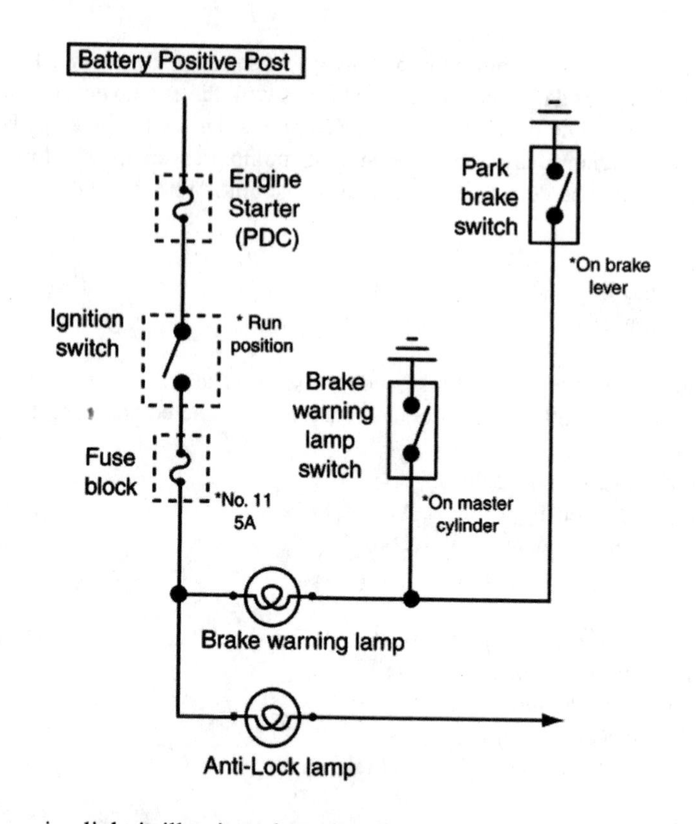

55. The brake warning light is illuminated continually with the ignition switch on in the above figure. All of these defects may be the problem **EXCEPT:**
    A. an open circuit in the wire to the parking brake switch.
    B. the wire to the parking brake switch touching the vehicle ground.
    C. a continually closed parking brake switch.
    D. low fluid level in the secondary section of the master cylinder reservoir. (E.5)

56. The special tool shown is used to:
    A. reset the brake warning light.
    B. hold the metering valve open during pressure bleeding.
    C. test the proportioning valve.
    D. adjust the bypass valve. (A.4.2)

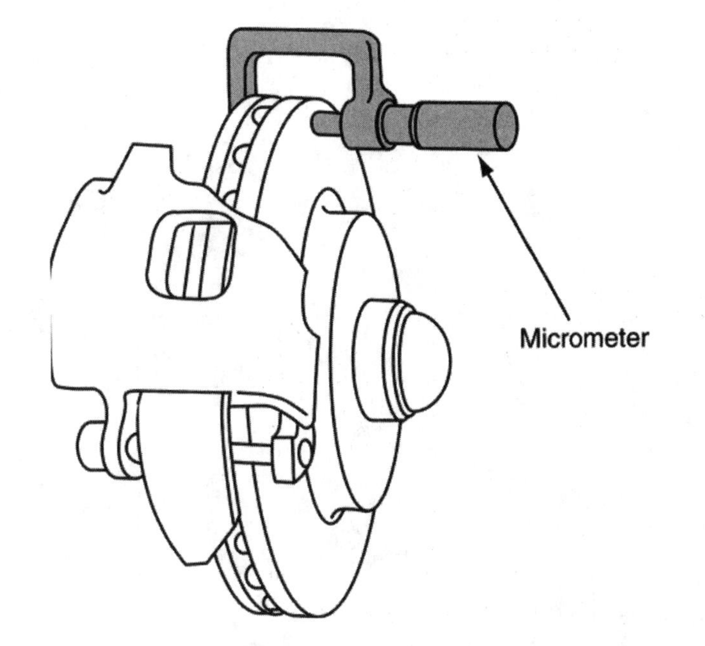

57. In the figure, Technician A says the brake rotor is being measured for runout. Technician B says this measurement should be made at three locations around the rotor. Who is right?
    A. A only
    B. B only
    C. Both A and B
    D. Neither A nor B (C.8)

58. The wiring harness from the right front ABS wheel speed sensor to the harness connector is badly damaged from corrosion—exposing the wires. Technician A says to carefully clean the damaged section and wrap the damaged area with black tape. Technician B says a badly corroded harness should be replaced. Who is right?
    A. A only
    B. B only
    C. Both A and B
    D. Neither A nor B (F.9)

59. While checking the vacuum supply to the power brake booster, Technician A says the vacuum gauge should be connected between the one-way check valve and the brake booster. Technician B says with the engine idling, the vacuum supplied to the brake booster should be 8 to 10 in.Hg (74.4 to 67.6 kPa absolute). Who is right?
    A. A only
    B. B only
    C. Both A and B
    D. Neither A nor B (D.2)

6 Additional Test Questions for Practice

Additional Test Questions

Please note the letter and number in parentheses following each question. They match the task in Section 4 that discusses the relevant subject matter. You may want to refer to the overview using the cross-referencing key to help with questions posing problems for you.

1. While diagnosing ABS, Technician A says the first thing to do is connect the test equipment to the data link connector. Technician B says the amber brake system light should not come on at all during a system self-test. Who is right?
   A. A only
   B. B only
   C. Both A and B
   D. Neither A nor B (F.3)

2. Technician A says linings that are covered with grease, oil, or brake fluid can be cleaned and reused. Technician B says it is only necessary to change the shoes on an assembly when they are worn to their specified limits. Who is right?
   A. A only
   B. B only
   C. Both A and B
   D. Neither A nor B (B.5)

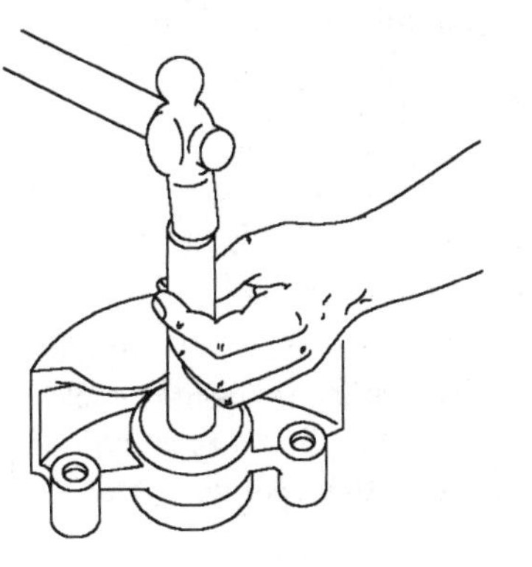

3. What operation is being performed in the figure?
   A. Installing a square-cut seal
   B. Installing a caliper piston
   C. Installing dust boot
   D. Honing the caliper bore (C.7)

4. The flexible wiring harness between the wheel sensor and the main harness that is mounted to the suspension is damaged. The wires have snapped. Technician A uses solderless crimp connector and heat shrink tubing to repair the wires. Technician B uses solder and heat shrink tubing to repair the wires. Who is right?
   A. A only
   B. B only
   C. Both A and B
   D. Neither A nor B (F.9)

5. A manual bleeding procedure is being done to a vehicle with disc/drum brakes. If air is suspected in the master cylinder bore, which should be bled first?
   A. The rear wheel cylinders
   B. The left front caliper
   C. The height-sensing proportioning valve
   D. The master cylinder (A.4.2)

6. Brake shoes are being replaced on a car. Technician A says to lubricate the backing plate ledges (bosses). Technician B says to groove the backing plate ledges (bosses) to match the shoe arc. Who is right?
   A. A only
   B. B only
   C. Both A and B
   D. Neither A nor B (B.6)

7. Technician A says the parking brake indicator light goes out when the ignition switch is turned off. Technician B says the parking brake indicator light may also be used in some vehicles to indicate problems in the antilock brake system. Who is right?
   A. A only
   B. B only
   C. Both A and B
   D. Neither A nor B (E.5)

8. Fluid drips on the carpet when the brake pedal is depressed on a vehicle with manual brakes. Which of the following is the cause?
   A. A defective master cylinder cup
   B. A torn input pushrod boot
   C. An overfilled master cylinder
   D. A leaking take-up valve (A.1.5)

9. During drum brake shoe installation, Technician A says the parking brake rear cable and conduit tension should be checked whenever the rear drum brakes are removed. Technician B says any grease will be suitable for lubricating the backing plate shoe pads. Who is right?
   A. A only
   B. B only
   C. Both A and B
   D. Neither A nor B (B.5 and B.6)

10. After the master cylinder is bled and installed, excessive brake pedal effort is required for brake application. Technician A says the vacuum hose connected to the brake booster may be restricted. Technician B says there may be insufficient clearance between pushrod and piston that allows pressure to build up in the system. Who is right?
    A. A only
    B. B only
    C. Both A and B
    D. Neither A nor B (A.1.6)

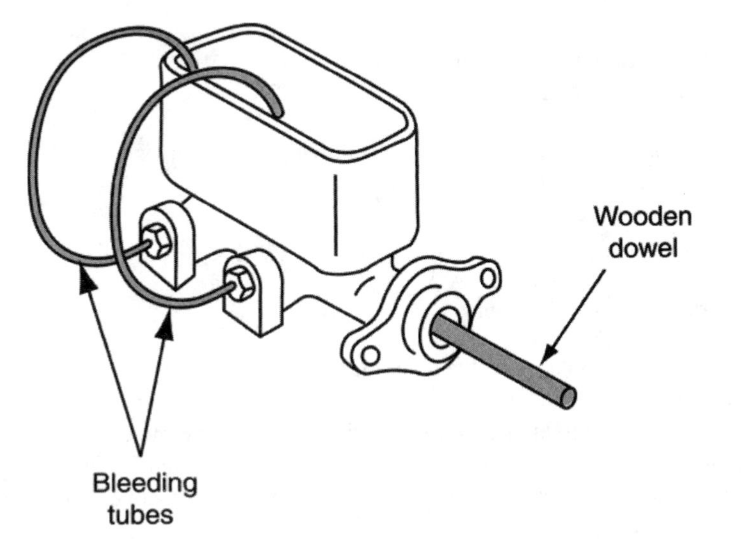

11. Refer to the figure for the following question: Technician A says that both reservoirs must be filled. Technician B says that repeatedly depressing the blunt end of the drift will remove air. Who is right?
    A. A only
    B. B only
    C. Both A and B
    D. Neither A nor B (A.1.6)

12. The brakes are applied on a vehicle with front disc/rear drums. The front brakes lock under pressure, but the rear brakes do not. Technician A says that seized wheel cylinder pistons can cause this. Technician B says a kinked brake line can cause this. Who is right?
    A. A only
    B. B only
    C. Both A and B
    D. Neither A nor B (B.1 and A.2.1)

13. A rotor surface needs to be made nondirectional after resurfacing. Technician A says to use a power tool and an aluminum oxide sanding disc. Technician B says to use an aluminum oxide sanding block. Who is right?
    A. A only
    B. B only
    C. Both A and B
    D. Neither A nor B (C.10)

14. When the vehicle is in motion, the wheel speed sensor of an antilock brake system will:
    A. send a varying AC voltage signal to the ABS controller.
    B. send a 5-volt supply signal to the ABS controller.
    C. complete a circuit to ground when a fault occurs.
    D. send a modified digital square wave signal to the ABS controller. (F.7)

15. Excessive pedal effort is needed to stop a vehicle with front disc/rear drums. The brake pedal is not spongy and does not sink. This could be caused by:
    A. glazed front pads.
    B. air in the system.
    C. incorrect rear brake shoe return springs.
    D. a caliper piston sticking. (C.1 and C.2)

16. A vehicle has drum brakes. Which of the following conditions is LEAST Likely to cause vibration (chatter) during braking?
    A. Out-of-round drum
    B. Hard spots on drum
    C. A loose backing plate
    D. A torn wheel cylinder boot (B.2)

17. A three-function combination valve combines the functions of:
    A. the brake failure light switch, the proportioning valve, and the metering valve.
    B. the brake failure light switch, the fast-fill valve, and the metering valve.
    C. the load-sensing proportioning valve, the fast-fill valve, and the metering valve.
    D. the load-sensing proportioning valve, the proportioning valve, and the fast-fill valve. (A.3.2)

18. All of the following could cause the brake lights on a car to be inoperative **EXCEPT:**
    A. a bad ground at the sockets.
    B. a bad flasher.
    C. a bad directional switch.
    D. a bad wire. (E.6)

19. All of the following are true about rear drum brakes **EXCEPT:**
    A. replace worn springs and hardware.
    B. clean and lubricate self-adjuster star wheels.
    C. shoes soaked with brake fluid must be cleaned before reinstallation.
    D. cracked shoes indicate overheating. (B.5)

20. A drum brake is dragging; you should look for all of the following **EXCEPT:**
    A. contaminated brake fluid.
    B. mismatched tire sizes.
    C. plugged compensating ports in the master cylinder.
    D. sticking wheel cylinder pistons. (B.1)

21. Technician A says to replace the sealing washers when replacing the brake hose. Technician B says to replace the sealing washers each time the brake hose is removed from the caliper. Who is right?
    A. A only
    B. B only
    C. Both A and B
    D. Neither A nor B (A.2.4)

22. Technician A says a proportioning valve is installed in the hydraulic line to the front brakes. Technician B says a metering valve is used with disc/drum brake systems to keep the front discs from operating until the rear drums have started to work. Who is right?
    A. A only
    B. B only
    C. Both A and B
    D. Neither A nor B (A.3.2)

23. A blue coloring of a brake shoe indicates:
    A. fluid contamination.
    B. a brake drum that is beyond its wear limits.
    C. a normal wear condition.
    D. overheating. (B.3)

24. The rear wheels on a car with a disc/drum system lock up during heavy braking. This could be caused by:
    A. a bad proportioning valve.
    B. a bad pressure differential valve.
    C. a bad metering valve.
    D. leaking axle seals. (A.3.2)

25. Technician A says plastic-coated, parking brake cables need periodic lubrication. Technician B says damage to the plastic coating will impair the smooth operation of the system. Who is right?
    A. A only
    B. B only
    C. Both A and B
    D. Neither A nor B (E.3)

26. The front end of a vehicle dips too much while braking lightly. Which of the following would be the cause?
    A. A bad master cylinder
    B. A bad metering valve
    C. A bad proportioning valve
    D. A bad vacuum check valve (A.3.2)

27. When flushing a hydraulic system, Technician A says each bleeder screw is opened one at a time and bled until all of the contaminated fluid is removed. Technician B says the fluid may be reused if it appears to be clean. Who is right?
    A. A only
    B. B only
    C. Both A and B
    D. Neither A nor B (A.4.2)

28. Technician A says there are two types of brake line flares: double and single. Technician B says there is only one acceptable type of tubing flare: the ISO flare. Who is right?
    A. A only
    B. B only
    C. Both A and B
    D. Neither A nor B (A.2.4)

29. The hydroboost unit above is being replaced. All of the following procedures should be performed **EXCEPT:**
    A. bleeding the power steering system and hydroboost unit.
    B. inspecting all fittings and hoses for leaks.
    C. replacing power steering hoses to hydroboost unit.
    D. road testing for proper steering performance and brake action. (D.4)

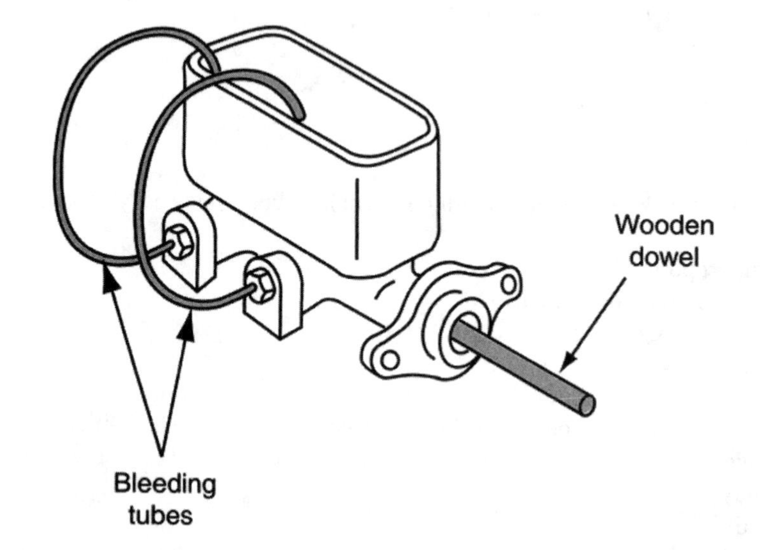

30. When bench bleeding a master cylinder using brake tubing, it is important that the ends of the tubes are submerged as shown in the figure because it:
    A. allows you to retain the expelled fluid.
    B. keeps the master cylinder cups from swelling.
    C. prevents contamination of the brake fluid.
    D. prevents air from being sucked back into the master cylinder. (A.1.6)

31. When diagnosing a vehicle with all brake light bulbs inoperative, the technician should start with the:
    A. bulbs in the stop lights.
    B. wiring to the switch.
    C. light sockets.
    D. fuse to the circuit. (E.6)

32. A customer is concerned that the brake pedal is easily depressed with little or no effect. Technician A says there could be air in the system and it should be bled. Technician B says it could be a leaking primary cup in the master cylinder. Who is right?
    A. A only
    B. B only
    C. Both A and B
    D. Neither A nor B (A.1.4)

33. Technician A says that for proper final brake finish, rotor machining must be done in one pass. Technician B says the rotor should be sanded to assist seating of the brake pads. Who is right?
    A. A only
    B. B only
    C. Both A and B
    D. Neither A nor B (C.10)

34. A vehicle equipped with a disc/drum system has brake pedal pulsation. This could be caused by:
    A. incorrect tightening of the wheel lug nuts.
    B. bulging flex-line hoses.
    C. a sticking master cylinder secondary piston.
    D. the installation of pads that are too hard. (C.2)

35. The rear wheels lock up too quickly during heavy braking on a vehicle with front disc/rear drum brakes. Which of these is the MOST-Likely cause?
    A. Air trapped in the brake lines
    B. Faulty metering valve
    C. Defective brake booster
    D. Malfunctioning proportioning valve (A.3.1 and A.3.2)

36. After the brakes have been bled and the system refilled, the next thing that should be done is:
    A. back off the drum brake star adjusters one full turn to avoid brake drag.
    B. road test the vehicle.
    C. pump the brakes several times and check for leaks.
    D. move the vehicle back and forth several times applying the pedal hard to set the brake self-adjusters. (A.4.2)

37. A vehicle with a hydroboost system has no power brake assist after the engine is shut off. This could be caused by:
    A. a defective accumulator.
    B. glazed brake pads.
    C. a sticking parking brake.
    D. incorrectly adjusted rear brakes. (D.4)

38. A master cylinder assembly is being removed from a vehicle with power assist. The technician should do all of the following **EXCEPT:**
    A. remove fluid from reservoir.
    B. plug the end of each brake line.
    C. use a fender cover to protect paint.
    D. remove the brake pedal pushrod. (A.1.6)

39. Brake pedal pulsation may be caused by:
    A. cone-shaped brake drums.
    B. brake shoes that are worn beyond specifications.
    C. low brake fluid.
    D. warped or distorted brake drums. (B.2)

40. A pressure gauge and adapter has been installed between the accumulator and the accumulator fluid outlet on a vehicle equipped with ABS. The vehicle is then started. After the pump stops, the engine is stopped and a reading is taken and found to be below specifications. All of the following components should be checked **EXCEPT:**
    A. the pump relay.
    B. the pump.
    C. the pressure switch.
    D. the accumulator.
    (F.4)

41. A restricted disc brake caliper hose could result in all of the following **EXCEPT:**
    A. dragging brakes.
    B. the vehicle pulling when braking.
    C. wheel bearing failure.
    D. poor stopping.
    (A.2.1 and A.2.2)

42. Self-adjusting drum brakes automatically:
    A. adjust lining-to-drum clearances.
    B. remove the slack from the parking brake cables.
    C. compensate for a dragging drum brake.
    D. adjust brake pad-to-rotor clearance.
    (B.8)

43. When reassembling a brake caliper,:
    A. the boot should be installed and seated, followed by the seal.
    B. coat the piston seal with clean brake fluid.
    C. leave the piston dry and install it through the boot and seal until it bottoms.
    D. plug the bleeder screw hole and the high-pressure inlet while installing the piston. (C.7)

44. Technician A says to replace the guide pins when rebuilding a caliper. Technician B says to replace the piston seal when rebuilding a caliper. Who is right?
    A. A only
    B. B only
    C. Both A and B
    D. Neither A nor B
    (C.7)

45. During a surge bleeding procedure:
    A. the end of the bleeder hose must be kept above the level of the fluid in the container.
    B. pump the pedal quickly several times with the bleeder screw closed.
    C. pump the pedal quickly several times while opening the bleeder screw.
    D. decrease the pressure bleeder chamber pressure to 10 psi (69 kPa). (A.4.2)

46. Technician A says a brake rotor should be refinished if it fails a lateral runout check. Technician B says a brake rotor should be refinished if it fails a thickness variation check. Who is right?
    A. A only
    B. B only
    C. Both A and B
    D. Neither A nor B
    (C.8)

47. With the engine idling, the vacuum at the brake booster is found to be below specification. The technician should then:
    A. connect the vacuum gauge directly to the intake manifold.
    B. conclude that the brake booster is faulty and rebuild it.
    C. check the brake fluid level.
    D. increase the engine rpm and recheck the vacuum.
    (D.3)

48. During installation, the brake shoes should be:
    A. sanded and cleaned with compressed air.
    B. retracted fully to allow the drum plenty of clearance during installation.
    C. adjusted to slightly less than the inside diameter of the drum.
    D. lubricated to aid in drum installation. (B.6 and B.8)

49. A constant pressure on the brake pedal results in the pedal sinking to the floor. No fluid leaks on external brake components can be seen. This could be caused by:
    A. an internal master cylinder leak.
    B. a crack on the master cylinder cap.
    C. a proportioning valve malfunction.
    D. a metering valve malfunction. (A.1.2)

50. Technician A says leaking or collapsed vacuum lines may cause weak power brake booster operation. Technician B says weak power brake booster operation may be caused by insufficient manifold vacuum. Who is right?
    A. A only
    B. B only
    C. Both A and B
    D. Neither A nor B (D.3)

51. Technician A says after a bearing is repacked, it must be reinstalled in the same race. Technician B says that it is acceptable to spin dry currently available bearings because they are superior to previous designs. Who is right?
    A. A only
    B. B only
    C. Both A and B
    D. Neither A nor B (E.2)

52. The front rotors are removed from a vehicle. The surface face on both rotors has small cracks and appears to have a blued finish. The brake rotors should be:
    A. machined with a non-directional finish.
    B. checked for lateral runout.
    C. checked for rotor parallelism.
    D. discarded. (C.8)

53. A vehicle has no master cylinder pushrod clearance. This problem may cause:
    A. a low brake pedal.
    B. pressure build-up and dragging brakes.
    C. a spongy brake pedal.
    D. fluid leaking past the primary piston cups. (A.1.3)

54. Which of the following will **NOT** cause the brake warning light to come on?
    A. A shorted light circuit
    B. A master cylinder leaking internally
    C. A leak in a brake line
    D. A torn caliper dust boot (E.5)

55. After the DTCs have been retrieved for the ABS (and the problem has been fixed), the technician should shut off the engine and then:
    A. erase the DTCs following manufacturers' procedure.
    B. fill the master cylinder reservoir to specifications.
    C. disconnect the brake booster vacuum hose to prevent accidental charging of the brake system.
    D. set the parking brake. (F.4)

56. Technician A says the brake shoes should be properly adjusted before the parking brake adjustment. Technician B says after the cable adjusting nut is tightened so the rear wheels have a slight drag, loosen the nut four turns. Who is right?
    A. A only
    B. B only
    C. Both A and B
    D. Neither A nor B (E.3 and E.4)

57. As shown, the vacuum booster pushrod length must be checked during all of the following repairs **EXCEPT:**
    A. brake shoe replacement.
    B. original assembly at the factory.
    C. master cylinder service.
    D. vacuum booster service. (A.1.3)

58. While testing the vacuum brake booster, the engine is operated for two minutes and then shut off. The brake pedal is then applied normally two to three times; the pedal goes down normally each time. Technician A says this indicates the booster is operating normally. Technician B says the one-way check valve is defective and should be replaced. Who is right?
    A. A only
    B. B only
    C. Both A and B
    D. Neither A nor B (D.1)

59. On an integral ABS with a high-pressure accumulator, Technician A says the master cylinder fluid level should be checked with a discharged accumulator. Technician B says the master cylinder fluid level should be checked with the engine running and the brake pedal applied. Who is right?
    A. A only
    B. B only
    C. Both A and B
    D. Neither A nor B (F.6)

60. With the engine running and the brake pedal applied and held, the brake pedal slowly moves downward. Technician A says brake fluid may be leaking past the master cylinder piston cups. Technician B says brake fluid may be leaking from a brake hose at one of the front wheels. Who is right?
    A. A only
    B. B only
    C. Both A and B
    D. Neither A nor B (A.1.4 and A.2.3)

Appendices

Answers to the Test Questions for the Sample Test Section 5

1. A	16. D	31. A	46. A
2. B	17. A	32. B	47. B
3. B	18. A	33. C	48. B
4. A	19. B	34. B	49. C
5. A	20. D	35. B	50. A
6. B	21. A	36. C	51. B
7. B	22. C	37. B	52. D
8. C	23. A	38. C	53. A
9. D	24. C	39. A	54. C
10. D	25. D	40. D	55. A
11. B	26. A	41. A	56. B
12. D	27. C	42. B	57. D
13. A	28. C	43. C	58. B
14. C	29. D	44. C	59. A
15. D	30. C	45. B	

Explanations to the Answers for the Sample Test Section 5

Question #1

Answer A is correct. Only Technician A is correct. The tool is being used to match the width of the brake shoes to the inside diameter of the brake drum before drum installation.

Answer B is wrong. The tool being used in the picture is not used to measure brake lining wear. A drum micrometer is used to check the drum diameter.

Answer C is wrong. Only Technician A is correct.

Answer D is wrong. Only Technician A is correct.

Question #2

Answer A is wrong. Answer A is a true statement but is not the correct answer because this is an EXCEPT question. The nuts securing the master cylinder to the booster will have to be removed.

Answer B is correct. The ends of the brake lines do not have to be plugged when the master cylinder is removed. However, it is a good practice to cap the threaded openings in the master cylinder to prevent fluid leakage while transporting the master cylinder to the workbench.

Answer C is wrong. Answer C is a true statement but is not the correct answer this is an EXCEPT question. The float level sensor will have to be disconnected when removing the master cylinder.

Answer D is wrong. Answer D is a true statement but is not the correct answer because this is an EXCEPT question. It is advisable to use a suction device to remove as much fluid as possible from the reservoir before removing the unit from the vehicle.

Question #3

Answer A is wrong. If the bearings were being replaced, then the races should also be replaced. Bearings and races are purchased in matched sets.

Answer B is correct. Only Technician B is correct. The spindle shaft should be inspected closely for wear during service of the bearings.

Answer C is wrong. Only Technician B is correct.

Answer D is wrong. Only Technician B is correct.

Question #4

Answer A is correct. Only Technician A is correct. A restricted brake hose will cause a decrease in fluid pressure to the affected caliper. As a result, the caliper will not be able to apply the same amount of force the other front caliper is exerting.

Answer B is wrong because air in the brake lines would cause a low and spongy pedal but would not cause a pulling problem.

Answer C is wrong. Only Technician A is correct.

Answer D is wrong. Only Technician A is correct.

Question #5

Answer A is correct. Only Technician A is correct. The only time the brake system needs to be bled is when the hydraulic system is opened or when air is present in the system. Many brake system repairs are done without the need to bleed the system afterwards.

Answer B is wrong. Many brake repairs can be made without coming into contact with the brake fluid.

Answer C is wrong. Only Technician A is correct.

Answer D is wrong. Only Technician A is correct.

Question #6

Answer A is wrong. Brake grabbing is not usually caused by a bent backing plate. This problem normally causes brake squeal or other noises.

Answer B is correct. Only Technician B is correct. Brake chatter could be caused by a loose anchor bolt. Other possible causes of brake chatter include misadjusted brake shoes, contaminated brake linings, misshapen drums, or loose wheel bearings.

Answer C is wrong. Only Technician B is correct.

Answer D is wrong. Only Technician B is correct.

Question #7

Answer A is wrong. Answer A is a true statement but is not the correct answer because this is an EXCEPT question. It is a good practice to clean the brake dust a residue from the drum before measuring it.

Answer B is correct. In this except-type question, all of the statements are true except B. If the inside diameter of a drum is less than specifications, it should not be replaced. In fact, it is fine providing the shoes fit inside the diameter.

Answer C is wrong. Answer C is a true statement but is not the correct answer because this is an EXCEPT question. It is wise to measure the drum at multiple locations to check for the drum being out-of-round.

Answer D is wrong. Answer D is a true statement but is not the correct answer because this is an EXCEPT question. The drum should always be checked for cracks, hard spots and scoring.

Question #8

Answer A is wrong. Technician B is also correct.

Answer B is wrong. Technician A is also correct.

Answer C is correct. Both Technicians are correct. Petroleum contaminated brake fluid can damage any part containing rubber including the master cylinder, calipers, wheel cylinders, hoses, and valves. Replacement of the affected components and brake flushing is the only way to insure proper brake function.

Answer D is wrong. Both Technicians are correct.

Question #9

Answer A is wrong. Cleaning solvent will leave a residue on the bores that is undesirable.

Answer B is wrong. Oil soap will leave a petroleum film on the bore that will cause the piston seals to swell.

Answer C is wrong. Using hot water will not clean the bores well enough and also would promote rust..

Answer D is correct. The bores of a wheel cylinder should only be cleaned with denatured or isopropyl alcohol. Other cleaning agents may leave a film behind, which can contaminate the fluid/

Question #10

Answer A is wrong. Answer A is a true statement but is not the correct answer because this is an EXCEPT question. Some caliper components and hoses are reverse threaded.

Answer B is wrong. Answer B is a true statement but is not the correct answer because this is an EXCEPT question. Most hoses are attached to the chassis and secured with a hose lock clip.

Answer C is wrong. Answer C is a true statement but is not the correct answer because this is an EXCEPT question. It is a good practice to clean the ends of the hoses prior to removal to reduce the chances of getting contamination in the system.

Answer D is correct. Replacement hoses should always match the length of the original equipment. If longer hoses were used, there would be an increased chance of the hose rubbing on a suspension component or possibly getting crimped because of routing problems.

Question #11

Answer A is wrong. Fluid leaking past the master cylinder cups would cause a spongy and possibly a sinking pedal. This action is sometimes referred to as the master cylinder bypassing fluid.

Answer B is correct. Only Technician B is correct. A misadjusted stoplight or cruise control disable switch could cause the brakes to drag by preventing full release of the brakes.

Answer C is wrong. Only Technician B is correct.

Answer D is wrong. Only Technician B is correct.

Question #12
Answer A is wrong. Wheel bearing adjustment would not cause a spongy pedal. A spongy pedal is usually caused by air in the brake system.
Answer B is wrong. Poor wheel bearing adjustment would not cause a hard pedal. A hard pedal is usually caused by a lack of power assist or having glazed pads.
Answer C is wrong. Poor wheel bearing adjustment would not cause a rising pedal.
Answer D is correct. Excessive end play would cause brake pedal pulsations. One of the causes of excessive end play is loose or damaged wheel bearings.

Question #13
Answer A is correct. Only Technician A is correct. Answer A is the most correct choice. On an ABS with a high-pressure accumulator, the accumulator is typically depressurized by pumping the brake pedal 25 times with the ignition switch off. Loosening or opening any part of the hydraulic system can be dangerous because the fluid is under the high pressure of the accumulator.
Answer B is wrong. Only Technician A is correct. Pumping the brake pedal with the key on allows the pump to run and build pressure back up to the operating pressure, which on some systems could be as high as 2,000 psi.
Answer C is wrong. Only Technician A is correct.
Answer D is wrong. Only Technician A is correct.

Question #14
Answer A is wrong. Adjusting the brake shoes until there is a slight drag on the wheel would cause the shoes to overheat and wear prematurely.
Answer B is wrong. Adjusting the brake shoes until there is a slight drag on the wheel would cause the shoes to overheat and wear prematurely.
Answer C is correct. When making final adjustments on the drum brakes, the star adjuster should be adjusted until there is a slight drag on the drum and then backed off until the wheel spins freely.
Answer D is wrong. This method does not fully describe the procedure.

Question #15
Answer A is wrong. Having air anywhere in the brake system will cause a spongy brake pedal.
Answer B is wrong. A faulty metering valve would cause the front end to nose-dive along with premature wear of the front brake linings.
Answer C is wrong because a defective brake booster would cause a hard pedal problem.
Answer D is correct. Trapped air in the master cylinder will cause a low, spongy pedal. A faulty metering valve will cause nose-diving and front wheel lock-up, not rear wheel lock-up. A faulty booster will MOST-Likely cause excessive pedal effort. The proportioning valve regulates pressure to the rear brakes under heavy brake to prevent rear wheel lock-up.

Question #16
Answer A is wrong. A faulty proportioning valve would affect rear brake application.
Answer B is wrong. A missing sealing washer would cause a brake fluid leak.
Answer C is wrong. A seized equalizer would prevent proper parking brake adjustment.
Answer D is correct. A restricted brake hose may force hydraulic fluid pressure through the flex hose and apply the brake, but may retain the fluid pressure in the caliper when the brake pedal is released. A way to check for this problem would be to crack the bleeder valve on the suspected wheel and look for residual pressure trapped in the caliper.

Question #17
Answer A is correct. Only Technician A is correct. All bench bleeding accomplishes is the removal of air from the master cylinder. Therefore it is possible that the system has air in it. In fact, it probably does even though the lines were capped. It is impossible to disconnect the lines and avoid letting some air in.
Answer B is wrong. Any time that the system is opened up, there is a chance for air to enter the system.
Answer C is wrong. Only Technician A is correct.
Answer D is wrong. Only Technician A is correct.

Question #18
Answer A is correct. Worn caliper pins and bushings could cause the caliper not to slide properly. On a sliding or floating caliper, the housing has to be able to move when the piston pushes on the inside pad. This movement allows the outside pad to engage the rotor and help stop the vehicle.
Answer B is wrong. A leaking caliper piston seal would cause fluid loss in the system.
Answer C is wrong. A leaking brake hose would cause fluid loss in the system.
Answer D is wrong. Excessive lateral rotor run out would cause a pedal pulsation with the brakes applied.

Question #19
Answer A is wrong. The height sensing proportioning valve never regulates pressure to the front wheels.
Answer B is correct. Only Technician B is correct. The height sensing proportioning valve does regulate pressure to the rear wheels to help prevent lock-up during hard braking. Height-sensing proportioning valves operate differently than regular proportioning valves. When the vehicle is loaded, the height sensing proportioning valve allows more brake pressure to reach the rear wheels because of increased traction.
Answer C is wrong. Only Technician B is correct.
Answer D is wrong. Only Technician B is correct.

Question #20
Answer A is wrong. Rotors must be checked for minimum thickness, parallelism, and lateral runout. When machining rotors, a technician should try to only take the minimum amount of metal to produce a smooth, even finish.
Answer B is wrong. The rotor does not need to be removed and reseated after the first cut is made. Care should be taken during the installation of the rotor before beginning the first cut.
Answer C is wrong. Neither Technician is correct.
Answer D is correct. Neither Technician is correct. Rotors must be measured for minimum thickness, parallelism and lateral runout. Only remove enough metal stock from a rotor to produce an even, smooth surface. Always measure the rotor before and after machining. Discard a rotor if it is not within specification.

Question #21
Answer A is correct. A leaking master cylinder piston cup will cause an external leak at the rear of the master cylinder. Because brake fluid tends to destroy paint, the peeling paint could be a good indication of a brake fluid leak.
Answer B is wrong. Since the vacuum booster doesn't normally contain brake fluid, a diaphragm leak would only affect brake operation and would not cause a brake fluid leak.
Answer C is wrong. Incorrect master cylinder pushrod adjustment will not cause an external master cylinder leak.
Answer D is wrong. Fuel cannot be drawn into the booster because the engine makes vacuum not the booster.

Question #22
Answer A is wrong. Technician B is also correct.
Answer B is wrong. Technician A is also correct.
Answer C is correct. Both Technicians are correct. A high-pressure accumulator in an ABS vehicle must be discharged before a brake line or any brake component is disconnected. Some manufacturers recommend relieving the accumulator pressure before discarding the accumulator.
Answer D is wrong. Both Technicians are correct.

Question #23
Answer A is correct. Only Technician A is correct. Vacuum bleeding pulls fluid and air through the system. The key to bleeding this way is to make sure the master cylinder remains full.
Answer B is wrong. Vacuum bleeding does not add or push fluid into the system.
Answer C is wrong. Only Technician A is correct.
Answer D is wrong. Only Technician A is correct.

Question #24
Answer A is wrong. Technician B is also correct.
Answer B is wrong. Technician A is also correct.
Answer C is correct. Both Technicians are correct. While replacing the brake pads, a good technician always cleans and lubricates the caliper slide pins and bushings along with the caliper slides. A thorough inspection of the mounting hardware, the caliper assembly, and brake hoses should be performed. Any worn components found should be replaced.
Answer D is wrong. Both Technicians are correct.

Question #25
Answer A is wrong. There is no sign of any problem in the system. The light is working as designed.
Answer B is wrong. There is no sign of any problem in the system. The light is working as designed.
Answer C is wrong. Neither Technician is correct.
Answer D is correct. Neither Technician is correct. Whenever the engine is started on a vehicle with antilock brakes, the warning light will come for a few seconds and then go out. This is normal. If the antilock light stays on or comes on while driving, the antilock computer has detected a fault. At this point, testing is needed.

Question #26
Answer A is correct. A kinked brake line could cause no hydraulic pressure to the left caliper. To repair this problem, a technician should install a new piece of steel brake line in place of the kinked line. Never install copper line or use compression fittings to repair a failed brake line.
Answer B is wrong. A seized left-front caliper piston would not block the brake fluid from getting into the caliper.
Answer C is wrong. The right-front wheel is working normally.
Answer D is wrong. A binding pushrod would affect pressure to more than just one wheel.

Question #27
Answer A is wrong. The purpose of the fast-fill valve is to rapidly fill the spool area of the primary piston with fluid from the reservoir.
Answer B is wrong. The fast-fill valve is located between the primary piston and the reservoir.
Answer C is correct. The spool area of the primary piston is the area that is provided fluid by the fast-fill valve.
Answer D is wrong. Fluid is ready to be sent to the actual wheel cylinder and caliper. The fill process has already occurred.

Question #28
Answer A is wrong. This does not indicate a restriction in the power booster hose. A restriction would cause a hard pedal and extra effort to stop the vehicle.
Answer B is wrong. A defective check valve could cause no assisted brake applications if the engine died.
Answer C is correct. A common way to check the action of a vacuum power brake booster is to check the brake pedal movement when the engine off. With the engine off, pump the brake pedal several times and hold the pedal in the applied position. When the engine is started, the pedal should move slightly downward. This is normal and indicates a normal supply of vacuum to the booster.
Answer D is wrong. Low vacuum would not allow the booster to react and the pedal would stay in the same position whether the engine was off or on.

Question #29
Answer A is wrong. Compressed air should never be used around any brake component.
Answer B is wrong. Using a lint-free cloth will not thoroughly clean the drum surface.
Answer C is wrong. Tapping the drum with a hammer will not thoroughly clean the drum surface.
Answer D is correct. The best way to clean the metal residue from the drum surface is to wash it in warm soapy water and wipe it dry with a lint-free cloth. If the metal residue is left on the drum surface, it will get embedded into the brake shoe and cause a metallic rubbing sound when the brakes are applied.

Question #30
Answer A is wrong. The proportioning valve limits brake pressure to the rear wheels during hard braking to help prevent rear wheel lock-up. This valve does not increase pressure during moderate brake application.
Answer B is wrong. The proportioning valve limits brake pressure to the rear wheels during hard braking to help prevent rear wheel lock-up. The proportioning valve does not affect brake pressure during moderate brake applications.
Answer C is correct. In the figure, the pressure gauges are connected to the input side of the valve and to the output for the rear brakes. When the brakes are applied, the proportioning valve decreases the pressure to the rear brakes during hard brake applications. During the light applications, the pressure is unaffected by the valve.
Answer D is wrong. The proportioning valve limits brake pressure to the rear wheels during hard braking to help prevent rear wheel lock-up. The proportioning valve does not affect brake pressure at all during light brake applications.

Question #31
Answer A is correct. Insufficient master cylinder pushrod clearance could cause a rising pedal. The reason this would occur is that the compensating ports in the master cylinder would be blocked off and pressure would start to build as the fluid heated up.
Answer B is wrong. A master cylinder piston seal leak could cause either an external fluid leak or a soft and spongy brake pedal.
Answer C is wrong. Reversed brake shoes could cause the drum brakes to drag or stick.
Answer D is wrong. Sticking wheel cylinder pistons could cause poor drum brake stopping power or dragging brakes.

Question #32
Answer A is wrong because grease contamination on the left side linings would cause a brake pull to the right side.
Answer B is correct. The direction a vehicle pulls toward during the application of the brakes is the side of the vehicle that has the most braking power.
Answer C is wrong because swollen piston seals could cause leaks or sticking pistons in the master cylinder but would not cause a brake pull.
Answer D is wrong because a plugged compensating port could cause the brakes to drag or lock up.

Question #33
Answer A is wrong. Answer A is a true statement but is not the correct answer because this is an EXCPET question It is a good practice to attach a hose to the bleeder and submerge the opposite end in a container of brake fluid.
Answer B is wrong. Answer B is a true statement but is not the correct answer because this is an EXCEPT question. The brake pedal should be applied with moderate pressure while an assistant opens the bleeder screw.
Answer C is correct. All of the statements in this question are true or correct, except C. If the bleeder screw is left open when the pedal is released and moves up, air will be drawn through the valve and enter the system.
Answer D is wrong. Answer B is a true statement but is not the correct answer because this is an EXCEPT question. The bleeding process should continue until the fluid is free of air bubbles.

Question #34
Answer A is wrong. Compression fittings should never be used in any brake line repair. Another note of caution would be to always use steel tubing in performing brake line repairs.
Answer B is correct. Only Technician B is correct. Whenever a brake line is damaged, it should be replaced. If a small section of the line is replaced it must incorporate double flare or ISO fittings and couplers. Compression fittings do not allow for these. Whenever a brake line needs to be shaped or bent, a tool-bending tool should be used. Trying to form the line by hand may result in dents or kinks.
Answer C is wrong. Only Technician B is correct.
Answer D is wrong. Only Technician B is correct.

Question #35
Answer A is wrong. A quick take-up valve that opens too early would cause a low brake pedal.
Answer B is correct. An overfilled brake fluid reservoir may cause the brakes to drag. The reason this could happen is that as the brake fluid temperature rises, it expands and starts building a pressure in the system. This pressure could cause the brakes to self apply and drag.
Answer C is wrong. A faulty proportioning valve would cause a problem in the rear brakes.
Answer D is wrong. A blocked vacuum check valve would cause a hard brake pedal.

Question #36
Answer A is wrong. The picture shows a combination valve, which includes a metering valve, a pressure differential valve, and a proportioning valve. The gauges are installed in series with the lines going to the rear brakes, so the correct answer is that the proportioning valve is being tested.
Answer B is wrong. The master cylinder quick take-up valve is not shown in the picture. The picture shows a combination valve, which includes a metering valve, a pressure differential valve, and a proportioning valve.
Answer C is correct. The picture shows that the proportioning valve is being tested because the pressure gauges are installed in series with the lines going to the rear brakes.
Answer D is wrong. A caliper is not shown in the picture. The picture shows a combination valve, which includes a metering valve, a pressure differential valve, and a proportioning valve.

Question #37
Answer A is wrong. This would cause a leak around the master cylinder, not the caliper.
Answer B is correct. A bad caliper seal would cause fluid residue around the caliper bore area.
Answer C is wrong. Restricted brake lines would cause poor stopping or possibly a brake-pulling problem.
Answer D is wrong. Worn caliper pins and bushings would cause uneven brake wear or possibly a brake-pulling problem.

Question #38
Answer A is wrong Technician B is also correct.
Answer B is wrong. Technician A is also correct.
Answer C is correct. Both Technicians are correct. The shoe ledges on the backing plate should be lubricated and if they are dry, a squeaking noise may be heard during brake application. If the shoe ledges are slightly scored, they can be filed down and sanded smooth, then a fresh coat of lubricant should be applied.
Answer D is wrong. Both Technicians are correct.

Question #39
Answer A is correct. Only Technician A is correct. When a fault occurs in the antilock brake system, the controller has the ability to store an error code in its memory. The error code will direct you to the affected area, but not necessarily determine that a particular component is faulty. Only testing the circuit and components, following the service manual procedure, will reveal the malfunction. A wheel speed sensor can be tested using a digital storage scope.
Answer B is wrong. A component should never be replaced without first testing the component and related wiring for problems.
Answer C is wrong. Only Technician A is correct.
Answer D is wrong. Only Technician A is correct.

Question #40
Answer A is wrong. Performing all these steps would not be wise when there is a much easier way to test the brake warning light circuit.
Answer B is wrong. Performing this action would not cause the brake warning light to illuminate.
Answer C is wrong. A wise technician never replaces a component without testing it the correct way.
Answer D is correct. An easy way to check the operation of the brake warning light circuit is to disconnect the wire to the switch and temporarily ground it. If the circuit is fine, the lamp will light when grounded. This is what the switch should do when a differential failure occurs.

Question #41
Answer A is correct. Only Technician A is correct. Battery power is supplied to the brake light switch. When the pedal is depressed, the switch closes and completes the circuit to the rear brake light bulbs. We already know that the fuse is good. Checking for power to the switch and out of the switch is a proper test procedure. If there were a short to ground, the fuse would not test good.
Answer B is wrong. If there was a short to ground in the harness, then the fuse would blow and there would be no voltage present at the brake switch.
Answer C is wrong. Only Technician A is correct.
Answer D is wrong. Only Technician A is correct.

Question #42
Answer A is wrong. Plugged vents in the master cylinder cover would not cause a spongy brake pedal.
Answer B is correct. Only Technician B is correct. Air in the brake hydraulic system would cause a spongy pedal. The reason for this is that air can be compressed under pressure so the pedal would feel soft and spongy during brake application.
Answer C is wrong. Only Technician B is correct.
Answer D is wrong. Only Technician B is correct.

Question #43
Answer A is wrong. Technician B is also correct.
Answer B is wrong. Technician A is also correct.
Answer C is correct. Both Technicians are correct. It is normal, on some vehicles, for the brake pedal to pulsate when the ABS is working. Brake pedal pulsations can also be caused by excessive rotor runout.
Answer D is wrong. Both Technicians are correct.

Question #44
Answer A is wrong. The metering valve delays pressure to the front wheels until the rear brake shoes have a chance to apply.
Answer B is wrong. The valve that restricts brake pressure to the rear brakes under heavy braking is the proportioning valve.
Answer C is correct. A metering valve is used on systems with front disc and rear drum brakes. During brake application, the fluid pressure must overcome the force of the brake shoe return springs and force the shoes outward. A little time is required to apply the rear drum brakes; therefore, the front disc brakes (which apply quickly) are slightly delayed during light brake applications so that both the front and rear brakes can be applied at the same time.
Answer D is wrong. The metering valve does not delay brake pressure to the front wheels during heavy pressure. The valve delays pressure to the front wheels until the rear brake shoes have a chance to apply.

Question #45
Answer A is wrong. A worn outer constant velocity joint would not cause a brake pedal pulsation while braking. A worn CV joint will typically cause a noise and/or a vibration while making a turn.
Answer B is correct. Only Technician B is correct. A loose front wheel bearing is a possible cause for pedal pulsation during braking. A way to further diagnose which wheel bearing is causing the problem would be to use a pyrometer to measure the hub temperature after road-testing the vehicle. A worn or loose bearing will cause the hub temperature to elevate on the suspected wheel.
Answer C is wrong. Only Technician B is correct.
Answer D is wrong. Only Technician B is correct.

Question #46
Answer A is correct. Only Technician A is correct. The brake system warning light comes on when there is a pressure difference on one side of the pressure differential switch. A leak in the hydraulic system could cause this difference in pressure.
Answer B is wrong. Equal pressure on both sides of the pressure differential switch would not cause the brake warning light to illuminate.
Answer C is wrong. Only Technician A is correct.
Answer D is wrong. Only Technician A is correct.

Question #47
Answer A is wrong. A problem in an ABS circuit on the R/R wheel is not likely to cause a code for the L/R wheel.
Answer B is correct. A DTC that indicates a faulty L/R wheel speed sensor indicates the problem is in that circuit and that circuit alone. Only one answer choice is related to that wheel's circuit, choice B. A larger than normal tire or wheel would not set a code. It would, however, affect ABS operation since the larger tire would rotate at a slower speed. The other answer choices are not specific to the L/R wheel.
Answer C is wrong. A larger tire on the R/F wheel would not set this code.
Answer D is wrong. An open circuit in the electronic brake control would set a code relating to a different problem, not a L/R speed sensor.

Question #48
Answer A is wrong. Component number 17 is the parking brake strut. This strut pushes the leading brake shoe into the drum when the parking brake is applied. If this part were omitted, the parking brakes would not function properly.
Answer B is correct. Only Technician B is correct. Component number 8 is the self-adjusting actuator cable. If this part were omitted, the brake would not self-adjust.
Answer C is wrong. Only Technician B is correct.
Answer D is wrong. Only Technician B is correct.

Question #49
Answer A is wrong. Technician B is also correct.
Answer B is wrong. Technician A is also correct.
Answer C is correct. Both Technicians are correct. Each time the parking brake is applied the caliper piston mechanically rotates and clamps the pads against the rotor. A special tool is needed to retract the caliper piston back into the rear caliper during brake service. After the brake pads are replaced and the caliper is mounted back in place on both sides, the parking brake should be applied several times to make sure it is working properly.
Answer D is wrong. Both Technicians are correct.

Question #50
Answer A is correct. Only Technician A is correct. Excessive lug nut torque can cause excessive lateral run-out of the rotor. Having all of the lug nuts installed at the correct torque is crucial to preventing rotor problems. This is especially true when dealing with wheels that are made of aluminum and composite materials. These types of materials are desirable because of their lightweight and decorative designs. However, these materials are more susceptible to bending if they are improperly torqued.
Answer B is wrong. An impact wrench should never be used to tighten lug nuts. An impact wrench will cause uneven torque on the lug nuts.
Answer C is wrong. Only Technician A is correct.
Answer D is wrong. Only Technician A is correct.

Question #51
Answer A is wrong. A faulty booster check valve will affect booster operation but will not cause a hiss with the brakes applied.
Answer B is correct. Only Technician A is correct. A leak in the booster will cause a hissing sound with the brakes applied.
Answer C is wrong. Only Technician B is correct.
Answer D is wrong. Only Technician B is correct.

Question #52
Answer A is wrong. If the pump were defective, the power steering would also not function properly.
Answer B is wrong. The accumulator in the hydro-boost system is used as a reserve in the event of the engine dying. Power assist should be normal even with a faulty accumulator until the engine is shut down.
Answer C is wrong. Neither Technician is correct.
Answer D is correct. Neither Technician is correct. The most likely fault in this system is a defective hydroboost hydraulic booster.

Question #53

Answer A is correct. Binding parking brake cables could cause the parking brakes to stick in the applied position after releasing the lever. It is important to advise customers to use their parking brakes on a regular basis to prevent the cable from rusting up and binding.

Answer B is wrong. If the self-adjusting cable were missing, then the parking brake could not be applied at all.

Answer C is wrong. The wheel cylinders are not part of the parking brake system.

Answer D is wrong. The ABS accumulator is not part of the parking brake system.

Question #54

Answer A is wrong. Bleeding the front brakes has nothing to do with park brake adjustment.

Answer B is wrong. Checking the tire pressure has nothing to do with park brake adjustment.

Answer C is correct. Before adjusting the parking brakes, the entire system should be inspected. There is no need to try to adjust the cables if they are broken, damaged, or kinked. The rear brake shoes should also be checked and adjusted before adjusting the parking brake.

Answer D is wrong. The parking brake system is independent of the hydraulic service brake system. There would be no need to fill the master cylinder with brake fluid before adjusting the parking brakes.

Question #55

Answer A is correct. In this except-type question, only answer A is wrong. An open in either circuit would not provide the necessary ground to illuminate the lamp.

Answer B is wrong. Answer A is a true statement but is not the correct answer because this is an EXCEPT question. The wire to the parking brake switch touching the vehicle ground could cause the brake warning light to stay on all of the time.

Answer C is wrong. Answer C is a true statement but is not the correct answer because this is an EXCEPT question. A continually closed parking brake switch would cause the brake warning light to stay on all of the time.

Answer D is wrong. Answer D is a true statement but is not the correct answer because this is an EXCEPT question. Low fluid level in the reservoir would cause the brake warning light to stay on all of the time if the vehicle is equipped with a low fluid sensor.

Question #56

Answer A is wrong. The tool shown is used to keep the metering valve open during pressure bleeding.

Answer B is correct. The tool shown is used to keep the metering valve open during pressure bleeding.

Answer C is wrong. The tool shown is used to keep the metering valve open during pressure bleeding.

Answer D is wrong. The tool shown is used to keep the metering valve open during pressure bleeding.

Question #57

Answer A is wrong. The figure shows a micrometer measuring the thickness of the rotor. A dial indicator would be needed to check for rotor runout.

Answer B is wrong. The technician should take measurements in at least 8 locations (every 45 degrees) around the rotor.

Answer C is wrong. Neither Technician is correct.

Answer D is correct. Neither Technician is correct. A rotor should be checked for thickness and parallelism with a micrometer. The figure for this question shows a micrometer taking a thickness measurement. Overall thickness will determine if the rotor can be reused. Making this measurement every 45 degrees will determine if the rotor is parallel or not.

Question #58

Answer A is wrong. This repair method is not an acceptable way of performing any wire repair. This wire problem is beyond any kind of repair and should be replaced.

Answer B is correct. Only Technician B is correct. If the wiring harness is badly damaged from corrosion it is recommended the harness be replaced. This section of harness is very susceptible to having problems because of its location near the hub and wheel.

Answer C is wrong. Only Technician B is correct.

Answer D is wrong. Only Technician B is correct.

Question #59
Answer A is correct. Only Technician A is correct. To check the vacuum supply to the booster, the vacuum gauge should be installed between the one-way check valve and the brake booster. With the engine idling, the vacuum should be 16 to 20 in Hg (47.4 to 40.6 kPa).
Answer B is wrong. When the engine is idling, there should be 16 to 20 inches of Hg (47.4 to 40.6 kPa).
Answer C is wrong. Only Technician A is correct.
Answer D is wrong. Only Technician A is correct.

Answers to the Test Questions for the Additional Test Questions Section 6

1. D
2. D
3. C
4. D
5. D
6. A
7. C
8. A
9. A
10. C
11. C
12. C
13. C
14. A
15. A
16. D
17. A
18. B
19. C
20. B
21. C
22. B
23. D
24. A
25. B
26. B
27. A
28. D
29. C
30. D
31. D
32. C
33. B
34. A
35. D
36. C
37. A
38. D
39. D
40. A
41. C
42. A
43. B
44. B
45. B
46. C
47. A
48. C
49. A
50. C
51. A
52. D
53. B
54. D
55. A
56. A
57. A
58. A
59. D
60. C

Explanations to the Answers for the Additional Test Questions Section 6

Question #1
Answer A is wrong. The Technician should begin troubleshooting ABS concerns by inspecting the system for obvious problems such as low fluid level or inoperative brake lights.
Answer B is wrong. The amber ABS light is supposed to come on when the vehicle is first started and then go off after a few seconds.
Answer C is wrong. Neither Technician is incorrect.
Answer D is correct. Neither Technician is correct. Technician A is wrong. A visual inspection of the system, including checking all brake lights and fluid levels, is first before performing any scanner functions. Technician B is also wrong. The amber ABS light should prove out for approximately three to six seconds and then go out. If the amber ABS light does not go out, there is a fault in the system.

Question #2
Answer A is wrong. Any time brake linings get contaminated with oil or grease, they should be replaced.
Answer B is wrong. There are more reasons for changing the brake shoes than just being worn out. Other possible reasons for replacing the brake shoes could be having been soaked with a petroleum material, brake fluid, being overheated, or having excessive cracks in the lining.
Answer C is wrong. Neither Technician is correct.
Answer D is correct. Neither Technician is correct because brake shoes should never be reused if they have been contaminated. There are also several other conditions that would cause the shoes to need to be replaced. See answer B.

Question #3
Answer A is wrong. The tools shown in the picture are not needed to install a square cut seal. A piston seal is installed by hand after lubricating the seal with clean brake fluid.
Answer B is wrong. The tools shown in the picture are not used to install a caliper piston.
Answer C is correct. The figure shows the use of a seal driver. The part the seal is being driven into is a caliper. The only caliper seal that is driven in place is the dust boot.
Answer D is wrong. The tools shown in the picture are not used to hone a cylinder bore.

Question #4
Answer A is wrong. The wheel speed sensor jumper harnesses are not repairable and must be replaced if they are damaged. If there is wire damage in other parts of the main wiring harness, then the solderless crimp-and-seal connectors or solder and heat shrink tubing can be used because they provide a weather resistant repair.
Answer B is wrong. The wheel speed sensor jumper harnesses are not repairable and must be replaced if they are damaged. If there is wire damage in other parts of the main wiring harness, then the solderless crimp-and-seal connectors or solder and heat shrink tubing can be used because they provide a weather resistant repair.
Answer C is wrong. Neither Technician is correct.
Answer D is correct. Neither Technician is correct. Wheel speed sensor jumper harnesses are not repairable and must be replaced if they are damaged. If there is wire damage in other parts of the main wiring harness, then the solderless crimp-and-seal connectors or solder and heat shrink tubing can be used because they provide a weather resistant repair.

Question #5
Answer A is wrong. The rear wheel cylinders should not be bled first. The master cylinder should be bled before the rest of the system.
Answer B is wrong. The left front caliper should not be bled first when air in the master cylinder is suspected.
Answer C is wrong. The height sensing proportioning valve should not be bled first.
Answer D is correct. When air is suspected to be trapped in the master cylinder, the master cylinder should be bled first.

Question #6

Answer A is correct. Only Technician A is correct. The ledges on the backing plate should be lubricated with some type of high-temperature grease. This allows the shoes to move freely when applied and released.
Answer B is wrong. The backing plate ledges should match the arc of the brake shoe. If they do not match, then the shoes are not right for the vehicle.
Answer C is wrong. Only Technician A is correct.
Answer C is wrong. Only Technician A is correct.

Question #7

Answer A is wrong. Technician B is also correct.
Answer B is wrong. Technician A is also correct.
Answer C is correct. Both Technicians are correct. The red brake warning light comes on for a few seconds when the ignition key is turned on and is turned off when the ignition is turned off regardless of conditions. The warning light informs the driver of brake system failures, such as the parking brake being on, uneven hydraulic pressures which may indicate a leak, and low ABS accumulator pressure.
Answer D is wrong. Both Technicians are correct.

Question #8

Answer A is correct. When brake fluid drips inside the vehicle, the fluid must be leaking from the rear of the master cylinder, as only this part is exposed to the interior of the vehicle. The pushrod boot is not designed to keep fluid in; it is designed to keep dirt out of the master cylinder. Both an overfilled master cylinder and leaking take up valve would show signs under the hood.
Answer B is wrong. The input pushrod boot is designed to keep dirt and contaminants out of the master cylinder. A torn seal would not cause an external leak at the master cylinder.
Answer C is wrong. An overfilled master cylinder would show leakage under the hood, not inside the vehicle.
Answer D is wrong. A leaking quick take-up valve would show leakage under the hood, not inside the vehicle.

Question #9

Answer A is correct. Only Technician A is correct. The parking brake assembly should be checked whenever the rear drum brakes are removed. If the technician finds a problem with the parking brake cable, a new cable can be easily installed while the brake shoes are disassembled.
Answer B is wrong. The plate shoes should only be lubed with high-temperature grease that won't dissolve under high heat conditions.
Answer C is wrong. Only Technician A is correct.
Answer D is wrong. Only Technician A is correct.

Question #10

Answer A is wrong. Technician B is also correct.
Answer B is wrong. Technician A is also correct.
Answer C is correct. Both Technicians are correct. Excessive pedal effort can be caused by a poor vacuum supply to the power brake booster. If the brake pushrod does not allow the master cylinder to return completely residual pressure will build up caused by the blocked port in the master cylinder. The brake pedal will get very hard as a result.
Answer D is wrong. Both Technicians are correct.

Question #11

Answer A is wrong. Technician B is also correct.
Answer B is wrong. Technician A is also correct.
Answer C is correct. Both Technicians are correct. During bench bleeding of a master cylinder, the reservoirs should be kept nearly full while a blunt object is moving the master cylinder's pistons in and out. This action will cause the air in the cylinder to be replaced by clean fluid.
Answer D is wrong. Both Technicians are correct.

Question #12
Answer A is wrong. Technician B is also correct.
Answer B is wrong. Technician A is also correct.
Answer C is correct. Both Technicians are correct. If the wheel cylinder pistons are seized, the rear brakes may not apply. Technician A is correct. A kinked brake line can block hydraulic pressure and prevent rear brake function. Technician B is also correct.
Answer D is wrong. Both Technicians are correct.

Question #13
Answer A is wrong. Technician B is also correct.
Answer B is wrong. Technician A is also correct.
Answer C is correct. Both Technicians are correct. To make the rotor nondirectional after resurfacing it, the rotor should be sanded. A finer grit is always used when a machine does the sanding.
Answer D is wrong. Both Technicians are correct.

Question #14
Answer A is correct. The wheel speed sensor consists of a toothed ring (also called a tone wheel) and a permanent magnet sensor. As the wheel turns the wheel speed sensor produces a varying AC voltage based on the changing magnetic field.
Answer B is wrong. The sensor receives a reference voltage from the controller that is used for diagnosis purposes within the controller.
Answer C is wrong. The speed sensor sends a varying AC voltage to the ABS controller based on wheel speed. If a fault occurs in the sensor, the ABS controller will see a change in the reference voltage signal it sends to the sensor.
Answer D is wrong. ABS speed sensors are PM generators that send a varying AC signal to the ABS controller based upon wheel speed.

Question #15
Answer A is correct. Glazed brake pads are an indication of heat-damaged linings, which can cause decreased braking ability and excessive pedal effort.
Answer B is wrong. Air in the brake system would cause a low and spongy pedal, not excessive pedal effort.
Answer C is wrong. Incorrect rear brake shoe return springs could not cause excessive pedal effort.
Answer D is wrong. A sticking caliper piston would cause a pull to one side while braking, not excessive pedal effort.

Question #16
Answer A is wrong. An out-of-round drum could cause brake vibration (chatter) during braking.
Answer B is wrong. Hard spots on the drum could cause brake vibration (chatter) during braking.
Answer C is wrong. A loose backing plate could cause brake vibration (chatter) while braking.
Answer D is correct. A torn wheel cylinder boot is the LEAST likely condition to cause brake vibration (chatter) during braking.

Question #17
Answer A is correct. The three-function combination valve combines the functions of the brake failure light switch, the proportioning valve, and the metering valve.
Answer B is wrong. The fast fill valve is part of the master cylinder, not the combination valve. The metering valve, the proportioning valve, and the brake failure switch (also called the pressure differential switch) are typically part the combination valve.
Answer C is wrong. The fast fill valve is part of the master cylinder, not the combination valve. Load-sensing proportioning valves are typically mounted independently at the rear of the vehicle, not as part of the combination valve. The metering valve, the proportioning valve, and the brake failure switch (also called the pressure differential switch) are typically part the combination valve.
Answer D is wrong. Load-sensing proportioning valves are typically mounted independently at the rear of the vehicle, not as part of the combination valve. The fast fill valve is part of the master cylinder, not the combination valve. The metering valve, the proportioning valve, and the brake failure switch (also called the pressure differential switch) are typically part the combination valve.

Question #18
Answer A is wrong. Answer A is a true statement but is not the correct statement because this is an EXCEPT question. A bad ground at the sockets could cause the lights to be inoperative.
Answer B is correct. A bad flasher is the least likely item to cause inoperative brake lights on a vehicle because the brake switch gets power from the fuse panel not the turn signal flasher. The flasher supplies power to the turn signal switch and causes the lights to flash when the turn signals are engaged.
Answer C is wrong. Answer C is a true statement but is not the correct answer because this is an EXCEPT question. A bad directional switch could cause the brake lights to be inoperative.
Answer D is wrong. Answer D is a true statement but is not the correct answer because this is an EXCEPT question. A bad wire could cause the brake lights to be inoperative.

Question #19
Answer A is wrong. Answer A is a true statement but is not the correct answer because this is an EXCEPT questions. Worn springs and hardware should always be replaced during brake service.
Answer B is wrong. Answer B is a true statement but is not the correct answer because this is an EXCEPT question. The self-adjuster should be cleaned and lubricated during brake service.
Answer C is correct. Always replace brake shoes that are contaminated with brake fluid or grease. Never reuse fluid soaked brake shoes. All other answers are correct.
Answer D is wrong. Answer B is a true statement but is not the correct answer because this is an EXCEPT question. Cracks in the lining material on the shoes indicate an overheating condition.

Question #20
Answer A is wrong. Answer A is a true statement but is not the correct answer because this is an EXCEPT question. Contaminated fluid could cause the brakes to drag by swelling the seals in the wheel cylinders.
Answer B is correct. Mismatched tire sizes will not cause brake drag.
Answer C is wrong. Answer C is a true statement but is not the correct answer because this is an EXCEPT question. Plugged compensating ports in the master cylinder could cause the brakes to drag by letting pressure build up in the hydraulic system.
Answer D is wrong. Answer D is a true statement but is not the correct answer because this is an EXCEPT question. Sticking wheel cylinder pistons could cause the brakes to drag by not releasing properly.

Question #21
Answer A is wrong. Technician B is also correct.
Answer B is wrong. Technician A is also correct.
Answer C is correct. Both Technicians are correct. Whenever a brake hose connection is broken loose, replace the sealing washer.
Answer D is wrong. Both Technicians are correct.

Question #22
Answer A is wrong. A proportioning valve is used on the rear wheels to help limit rear braking pressure during hard pedal applications. The reason that pressure is limited to the rear wheels under these conditions is because the forward weight shift causes more of the traction to be shifted to the front wheels. Thus, by limiting pressure to the rear wheels, the vehicle is less likely to go into a skid.
Answer B is correct. Only Technician B is correct. The metering valve is used on a vehicle with front disc/rear drum brakes. The metering valve holds off pressure to the front wheels until the rear wheel cylinders overcome the spring tension at the brake shoes. By keeping the front brakes from applying too early, the metering valve provides simultaneous application of the front and rear brakes.
Answer C is wrong. Only Technician B is correct.
Answer D is wrong. Only Technician B is correct.

Question #23

Answer A is wrong. Fluid contamination will cause an oil build up on the brake shoe.

Answer B is wrong. A brake drum beyond its wear limits will not cause an overheating condition.

Answer C is wrong. A normal brake drum wear condition will not cause the shoes to be blue.

Answer D is correct. A blue coloring on the metal part of the brake shoe indicates that the shoe has been overheated. If a technician finds this problem during an inspection, he/she should try to find and repair the root cause to the problem. The linings should also be replaced before returning the vehicle to service because they will likely have a glazed finish and not perform to meet specifications.

Question #24

Answer A is correct. The proportioning valve is designed to prevent rear-wheel lockup under heavy braking.

Answer B is wrong. A bad pressure differential valve could cause the brake indicator light to malfunction. The purpose of this valve is to alert the driver if there is unequal hydraulic pressure in the system.

Answer C is wrong. The main purpose of the metering valve is to hold off pressure to the front disc brakes until the rear drum brake shoes have a chance to apply. When working properly, the metering valve prevents nosedive in the front end while braking lightly.

Answer D is wrong. Leaking axle seals allow gear lube to soak the brake shoes thus causing them to lose friction.

Question #25

Answer A is wrong. Plastic-coated parking brake cables do not require periodic lubrication.

Answer B is correct. Only Technician B is correct. If the plastic coating gets damaged, the cable could malfunction.

Answer C is wrong. Only Technician B is correct.

Answer D is wrong. Only Technician B is correct.

Question #26

Answer A is wrong. A problem with the master cylinder would not likely cause this problem under light braking.

Answer B is correct. The main purpose of the metering valve is to hold off pressure to the front disc brakes until the rear drum brake shoes have a chance to apply. When working properly, the metering valve prevents nosedive in the front end while braking lightly.

Answer C is wrong. The proportioning valve limits pressure to the rear brakes under heavy braking pressure.

Answer D is wrong. A bad vacuum check-valve would cause a problem with the power assist system. If the check-valve got stopped up, there would be no power assist at any time. If the check-valve failed to open, there would be no power-assisted stops if the engine dies.

Question #27

Answer A is correct. Only Technician A is correct. To flush a brake system, a technician should begin by extracting as much of the old brake fluid from the reservoir as possible and then filling the reservoir with fresh, clean brake fluid. Next, each bleeder screw is opened one at a time and bled until all of the contaminated fluid is removed.

Answer B is wrong. A technician should never reuse old brake fluid for any reason. New, clean brake fluid from a sealed container should be used when adding brake fluid to any system.

Answer C is wrong. Only Technician A is correct.

Answer D is wrong. Only Technician A is correct.

Question #28

Answer A is wrong. The two types of brake flares are double flare and ISO flare. Single flares are never to be used on brake lines.

Answer B is wrong. There are two types of flares used on brake lines. The two types include the double flare and the ISO flare. Single flares are never to be used on brake lines.

Answer C is wrong. Neither Technician is correct.

Answer D is correct. Neither Technician is correct. There are two types of flares used on brake lines. The two types include the double flare and the ISO flare. Single flares are never to be used on brake lines.

Question #29
Answer A is wrong. Answer A is a true statement but is not the correct because this is an EXCEPT question. The power steering system and the hydroboost unit should be bled following the replacement of the hydroboost unit.
Answer B is wrong. Answer B is a true statement but is not the correct because this is an EXCEPT question. The fittings and hoses should be checked for leaks following the replacement of the hydroboost unit.
Answer C is correct. All procedures outlined are correct except C. It is not necessary to replace the power steering hoses, unless the hoses are found to be damaged or leaking.
Answer D is wrong. Answer D is a true statement but is not the correct because this is an EXCEPT question. The vehicle should be road-tested to check for proper power steering and brake operation.

Question #30
Answer A is wrong. Keeping the tubes submerged helps prevent air from being drawn in when the pistons are released.
Answer B is wrong. Keeping the tubes submerged helps prevent air from being drawn in when the pistons are released. The most common cause of swollen piston cups is someone accidentally adding the wrong kind of fluid such as engine oil or power steering fluid.
Answer C is wrong. Keeping the tubes submerged helps prevent air from being drawn in when the pistons are released. Keeping the tubes submerged does not affect contamination of the brake fluid.
Answer D is correct. The reason for making sure the tubes remain submerged during bench bleeding a master cylinder is simply to prevent the tubes from drawing in air as the pistons of the master cylinder are being moved.

Question #31
Answer A is wrong. It is very unlikely that every stoplight bulb could be faulty at the same time.
Answer B is wrong. The switch wiring is not a logical point to start diagnosing this problem. It is a good idea to start with the most basic items first, i.e., checking the fuse.
Answer C is wrong. The light sockets are not very likely to all be defective, causing all of the brake lights to be inoperative.
Answer D is correct. Inoperative stop lamps can be caused by several things. The best way to diagnose the problem is to think about it. If all the lamps do not light, it is very unlikely that the problem is burned out bulbs. The cause of the problem must be something that is common to all stop lamps. This is exactly what answer D suggests. The fuse is common to all lamps and would be the most logical place to begin diagnosis.

Question #32
Answer A is wrong. Technician B is also correct.
Answer B is wrong. Technician A is also correct.
Answer C is correct. Both Technicians are correct. Both of these problems could cause the concern and both should be checked. Air in the brake system would cause a soft and spongy brake pedal because air can be easily compressed. A leaking primary seal in the master cylinder could also cause a soft pedal with poor braking performance. If the master cylinder had this problem, it would need to be replaced.
Answer D is wrong. Both Technicians are correct.

Question #33
Answer A is wrong. It is sometimes necessary to make more than one final cut on a brake rotor.
Answer B is correct. Only Technician B is correct. Brake pads will seat more quickly and operate more quietly if the brake rotor is sanded after the final cut. Before installing the rotor on the vehicle, the rotor should be washed in warm soapy water to remove any metallic residue from its surface. This step helps prevent the brake linings from getting the metal flakes embedded into the pad material.
Answer C is wrong. Only Technician B is correct.
Answer D is wrong. Only Technician B is correct.

Question #34

Answer A is correct. A common cause of pedal pulsation is a distorted rotor and the primary cause of this distortion is over-tightening of lug nuts. The other answer choices would not cause pedal pulsation.

Answer B is wrong. Bulging flex hose could cause a pull while braking, but not a pedal pulsation.

Answer C is wrong. A sticking master cylinder secondary piston could cause the brakes to drag, but not a pedal pulsation.

Answer D is wrong. The wrong pads could cause the loss of the correct pedal feel, but not a pedal pulsation.

Question #35

Answer A is wrong. Air trapped in the brake lines would cause a soft and spongy brake pedal.

Answer B is wrong. The metering valve holds off pressure to the front wheels until the rear brakes have a chance to start applying.

Answer C is wrong. A faulty brake booster would likely cause excessive pedal effort.

Answer D is correct. The proportioning valve regulates pressure to the rear wheels during heavy braking to prevent the back wheels from locking up prematurely. A brake pressure test could be used to test the proportioning valve. The pressure to the rear wheels should be less than the pressure to the front wheels under heavy pedal application.

Question #36

Answer A is wrong. The brake shoes should have been adjusted prior to bleeding the brakes.

Answer B is wrong. The system needs to be inspected for leaks before taking it for a road test.

Answer C is correct. After cleaning all of the wheel areas with brake cleaner, the brake pedal should be pumped several times. The technician should then inspect the whole system for any signs of brake fluid leakage.

Answer D is wrong. The system needs to be checked for leaks before moving the vehicle.

Question #37

Answer A is correct. Reduced assist from a hydro-boost system after the engine is shut off is Most-Likely caused by a defective accumulator. The accumulator holds reserve fluid that, in the event of loss of fluid in the system, would assist in applying the brakes.

Answer B is wrong. Glazed pads would cause lack of braking power while driving not reduced power assist with the engine off.

Answer C is wrong. The parking brake system is operated using a system of cables and levers. It has nothing to do with the hydro-boost system.

Answer D is wrong. Incorrectly adjusted rear brake shoes would cause a low pedal reserve. This condition is not linked to the lack of power assist.

Question #38

Answer A is wrong. Answer A is a true statement but not the correct answer because this is an EXCEPT question.

Answer B is wrong. Answer A is a true statement but not the correct answer because this is an EXCEPT question.

Answer C is wrong. Answer C is a true statement but not the correct answer because this is an EXCEPT question.

Answer D is correct. This is an except-type question that has all true statements but one. That one false statement is answer D. When a master cylinder is being removed from a vehicle that has power brake assist, the brake pedal push rod does not need to be removed. This is also true for many non-power assisted systems.

Question #39

Answer A is wrong. Cone-shaped brake drums could cause inner or outer brake shoe wear but not pedal pulsation.

Answer B is wrong. Worn-out brake shoes could cause unusual noises but not pedal pulsation.

Answer C is wrong. Low brake fluid could cause a soft, spongy pedal.

Answer D is correct. Brake drums that are warped or distorted could cause brake pedal pulsation. Drums found to be warped or distorted should be measured with a drum micrometer to see if they can be machined on a lathe. The machine-to diameter should not be exceeded when cutting the drums on a brake lathe. A typical 9-inch drum will have a machine-to diameter of 9.060 inches and a discard diameter of 9.090 inches.

Question #40

Answer A is correct. The relay is the only component that can be ruled out as a possible cause for the low pressure. If the relay was bad, then the pump would not have turned on at all and there would be no pressure.

Answer B is wrong. Answer B is a true statement but is not the correct answer because this is an EXCEPT question. A faulty pump could cause the pressure to drop lower than specifications.

Answer C is wrong. Answer C is a true statement but is not the correct answer because this is an EXCEPT question. A faulty pressure switch could cause the pump to turn off too soon.

Answer D is wrong. Answer D is a true statement but is not the correct answer because this is an EXCEPT question. A faulty accumulator could cause the pressure in the system to be too low.

Question #41

Answer A is wrong. Answer A is a true statement but is not the correct answer because this is an EXCEPT question. A restricted brake hose could cause dragging brakes by not allowing the fluid to exhaust properly when the brake pedal is released.

Answer B is wrong. Answer A is a true statement but is not the correct answer because this is an EXCEPT question. A restricted brake hose could cause a vehicle pull during braking by restricting the flow of fluid in the affected circuit.

Answer C is correct. All of the statements are true except C. Wheel bearing failure will not result from brake hose restrictions.

Answer D is wrong. Answer D is a true statement but is not the correct answer because this is an EXCEPT question. A restricted brake hose could cause poor stopping by restricting the flow of fluid in the affected circuit.

Question #42

Answer A is correct. Self-adjusting drum brakes automatically adjust the lining-to-drum clearance. The action of stopping in reverse causes some parking brake adjusters to operate while other designs are designed to mechanically adjust whenever the parking brake is operated.

Answer B is wrong. The self-adjusters change the shoe-to-drum clearance instead of changing the parking brake cable adjustment.

Answer C is wrong. The self-adjusters will not compensate for a dragging drum brake. The adjuster can only move the shoes closer to the drum during normal operation.

Answer D is wrong. Drum brakes do not use a rotor.

Question #43

Answer A is wrong. The seal and piston should be installed before the boot.

Answer B is correct. While assembling a brake caliper, coat the piston seal with clean brake fluid. Do not attempt to install the seal after the boot has been installed. Also do not attempt to install a dry piston through the boot.

Answer C is wrong. The technician needs to coat the parts with clean brake fluid during assembly.

Answer D is wrong. If the bleeder screw hole is plugged while you attempt to install the piston, you will need to compress the air in the cylinder while pushing the piston in. An open bleeder screw allows the air to be pushed out of the way as the piston is being installed.

Question #44

Answer A is wrong. Caliper guide pins are not part of a normal caliper rebuilding kit.

Answer B is correct. Only Technician B is correct. The piston seal is always replaced when rebuilding a caliper.

Answer C is wrong. Only Technician B is correct.

Answer D is wrong. Only Technician B is correct.

Question #45

Answer A is wrong. The bleeder hose should be submerged into the brake fluid in the container to keep air from being drawn in through the bleeder.

Answer B is correct. Surge bleeding is a method in which the pedal is pumped hard several times with the bleeder closed. This rapid movement of the brake fluid is sometimes needed to pass all of the air bubbles out of the system.

Answer C is wrong. Pumping the pedal with the bleeder open would allow air to enter into the bleeder.

Answer D is wrong. This statement describes the pressure bleeding method.

Question #46
Answer A is wrong. Technician B is also correct.
Answer B is wrong. Technician A is also are correct.
Answer C is correct. Both Technicians are correct. A brake rotor should be refinished if it fails either the lateral run-out check or the thickness variation check.
Answer D is wrong. Both Technicians are correct.

Question #47
Answer A is correct. With the engine idling, if the vacuum from the intake manifold to the brake booster is found to be low, vacuum at the manifold should be measured before further testing and before replacing any parts.
Answer B is wrong. Thorough testing should always be performed before replacing any parts.
Answer C is wrong. The brake fluid level has nothing to do with low vacuum at the booster.
Answer D is wrong. The engine should supply adequate vacuum to the booster. Vacuum at idle should be approximately 16–18 in of Hg.

Question #48
Answer A is wrong. It is never acceptable to sand the linings or clean with compressed air. All brake linings need to be treated as if they contained asbestos. Asbestos is a cancer causing material than can get airborne for long periods of time. If a technician breathes air that contains asbestos, he runs a higher risk of developing lung cancer later in life.
Answer B is wrong. The shoes should be pre-adjusted using the measurement tool before installing the drum.
Answer C is correct. The brake shoes should be adjusted to slightly less than the inside diameter of the brake drum. Once the drum is installed, the technician can finish the adjustment procedure.
Answer D is wrong. Thee friction surface on the brake shoes should never be lubricated with anything.

Question #49
Answer A is correct. When constant pressure on the brake pedal results in the brake pedal moving to the floor, the problem is an internal leak in the master cylinder. This is answer A. The other answers would affect brake operation with a normal pedal feel.
Answer B is wrong. A cracked master cylinder cap would cause an external leak at the master cylinder.
Answer C is wrong. A proportioning valve malfunction would cause the rear brakes to be affected under heavy braking.
Answer D is wrong. The metering valve malfunction would cause the front brakes to be applied too soon under light braking.

Question #50
Answer A is wrong. Technician B is also correct.
Answer B is wrong. Technician A is also correct.
Answer C is correct. Both Technicians are correct. Weak power unit operation may be caused by leaking or collapsed vacuum lines and by insufficient manifold vacuum.
Answer D is wrong. Both Technicians are correct.

Question #51
Answer A is correct. Only Technician A is correct. After a bearing is repacked, it should be reinstalled in the same race.
Answer B is wrong. It is dangerous to use compressed air to spin dry bearings. This action could cause flying debris to injure the technician or someone nearby.
Answer C is wrong. Only Technician A is correct.
Answer D is wrong. Only Technician A is correct.

Question #52
Answer A is wrong. The rotors have been severely overheated and should be replaced.
Answer B is wrong. The rotors have been severely overheated and should be replaced.
Answer C is wrong. The rotors have been severely overheated and should be replaced.
Answer D is correct. Cracks on the face of the rotors along with a blued finish are an indication of severely overheated brake rotors. At this point the rotors should be discarded. Checking for lateral runout and parallelism are routine checks if the rotors are going to be cut. A non-directional finish is performed after a rotor is machined.

Question #53
Answer A is wrong because a lack of clearance could cause a high pedal instead of a low pedal.
Answer B is correct. Only Technician B is correct. Having no clearance between the pushrod and the master cylinder piston causes the compensating port to be blocked off This action could cause the brakes to drag or lock up.
Answer C is wrong. Only Technician B is correct.
Answer D is wrong. Only Technician B is correct.

Question #54
Answer A is wrong. Answer A is a true statement but is not the correct answer. A short to ground in the light circuit could cause the brake warning light to come on.
Answer B is wrong. Answer B is a true statement but is not the correct answer. A master cylinder with an internal leak could cause the brake warning light to come on.
Answer C is wrong. Answer C is a true statement but is not the correct answer. A leak in a brake line could cause the brake warning light to come on due to the pressure differential valve sensing an imbalance in the system.
Answer D is correct. All of the answers could cause the red brake warning light to illuminate except a torn caliper dust boot. This condition might cause future problems with the caliper but it would not cause the brake light to come immediately.

Question #55
Answer A is correct. After the DTCs have been retrieved for the ABS and the problem has been fixed, a technician should erase trouble codes. On some systems the DTCs can be erased through a scan tool. Other systems will reset during self-test.
Answer B is wrong. The information given with the question did not mention any fluid being lost during the repair.
Answer C is wrong. There is no reason to disconnect the vacuum brake hose after any ABS problem.
Answer D is wrong. There is no reason to set the parking brake after making a repair to the ABS system.

Question #56
Answer A is correct. Only Technician A is correct. The brake shoes should always be inspected and adjusted to specifications before adjusting the parking brake cable. This allows the technician to see the condition of the linings and to adjust the clearance between the linings and drums.
Answer B is wrong. This adjustment procedure would provide too much cable slack.
Answer C is wrong. Only Technician A is correct.
Answer D is wrong. Only Technician A is correct.

Question #57
Answer A is correct. The vacuum booster pushrod length must be checked during the original assembly at the factory by the manufacturer, master cylinder service, and vacuum booster service. It is not of concern when you are replacing brake shoes.
Answer B is wrong. Answer B is a true statement but is not the correct answer because this is an EXCEPT question. Pushrod length is checked during assembly at the factory.
Answer C is wrong. Answer C is a true statement but is not the correct answer because this is an EXCEPT question. Pushrod length is checked during master cylinder service.
Answer D is wrong. Answer D is a true statement but is not the correct answer because this is an EXCEPT question. Pushrod length should be checked during master cylinder service.

Question #58
Answer A is correct. Only Technician A is correct. If the booster is working properly, you should get 2 to 3 assisted brake applications after the engine is shut off. No assist (hard pedal) immediately after the engine is shut off indicates a faulty check valve or vacuum leak.
Answer B is wrong. The results of this test indicate the vacuum check valve is working as designed.
Answer C is wrong. Only Technician A is correct.
Answer D is wrong. Only Technician A is correct.

Question #59
Answer A is wrong. The brake fluid should be checked only after pumping the brake pedal 25 times with the ignition turned off. This process allows the brake fluid to be returned to the reservoir so the fluid level can be properly checked. This process lets the accumulator release the brake fluid it is storing; however, the accumulator is stilled considered to be charged.
Answer B is wrong. The brake fluid should be checked only after pumping the brake pedal 25 times with the ignition turned off. This process allows the brake fluid to be returned to the reservoir so the fluid level can be properly checked.
Answer C is wrong. Neither Technician is incorrect.
Answer D is correct. Neither Technician is correct. In an integral ABS with a high-pressure accumulator, the brake fluid level should be checked only after pumping the brake pedal 25 times with the ignition off. This process allows the brake fluid to be returned to the reservoir so the fluid can be checked accurately. The accumulator is still considered to be charged to approximately 500 psi and should be handled with care.

Question #60
Answer A is a wrong. Technician B is also correct.
Answer B is wrong. Technician A is also correct.
Answer C is correct. Both Technicians are correct. If brake fluid leaks past the primary cups of the master cylinder, the pedal will slowly move downward after the brakes have been applied. The leaking cups would prevent high pressure buildup in the master cylinder. The same would happen if there was a small leak in one of the brake hoses or lines.
Answer D is wrong. Both Technicians are correct.

Glossary

Acetone A highly flammable liquid sometimes used to clean parts.

Aeration To expose to the air or mix with air, as with a liquid; to charge a liquid with gas.

Accumulator In ABS brake systems, a chamber that stores fluid. In hydro-boost brake systems; a charged chamber used to assist braking if a failure occurs to the power steering system or if the engine stalls.

Air gap A small space between two parts.

Air shock A shock operating on the principles of air pressure; may also have a hydraulic section.

Alternating current An electric current whose polarity is constantly changing from positive to negative.

Antilock brakes (ABS) A computerized brake system that prevents wheel lock-up.

Asbestos Heat resistant material once used extensively in brake linings and clutch discs. Inhaling asbestos dust is harmful and may cause cancer.

Battery A device for storing electrical energy in chemical form.

Bench bleeding A procedure for bleeding air from a master cylinder before installing it in a vehicle.

Bleeder screw A small valve-like screw located on each wheel cylinder and master cylinder used to bleed air from the brake system.

Bleeding The act of removing air from a hydraulic brake system.

Boiling point The temperature at which a substance, such as a liquid, begins to boil.

Boot A flexible rubber or plastic cover over the end of wheel cylinders or master cylinder to keep out water and other foreign matter.

Bore May refer to the cylinder itself or to the diameter of the cylinder.

Brake A system used to slow or stop a vehicle; to slow or stop a vehicle.

Brake fade Loss of braking force due to heat buildup.

Brake light Red lamps at the rear of a vehicle to warn others that a braking action is taking place.

Brake light switch A switch found on the brake linkage or in the hydraulic system that activates the brake lights.

Brake line A small diameter rigid steel tube that connects the brake system to the brake hose which, in turn, connects to the wheel cylinders.

Brake pads A friction material applied to the disc by the caliper to slow or stop a vehicle.

Brake system The system in a vehicle that is used to slow or stop a vehicle.

Brake warning light An instrument panel lamp that warns of a brake system function or a malfunction.

Brinelling A condition in which a bearing or race has a series of dents or grooves worn into the surface.

Check valve A valve, usually spring loaded, that allows the passage of fluid or vapor in one direction but not the other.

Circuit A complete path for an electric current; a compete path for a fluid system.

Concentric grinding Grinding of the brake lining so it matches the brake drum.

Crocus cloth A fine abrasive cloth used for polishing or cleaning metal surfaces.

Cruise control A system that allows a vehicle to maintain a preset speed though the driver's foot is not on the accelerator pedal.

Denatured alcohol Ethyl alcohol used to clean brake parts.

Dial indicator A measuring tool used to determine end play motion.

DOT An abbreviation for the Department of Transportation.

Drum That part of a brake that rotates with the wheel and the brake shoes press against to slow or stop the vehicle.

Eccentric grinding Grinding portions of the brake lining to a different contour than that of the brake drum.

Electronic control unit A digital computer that controls engine and transmission functions.

Equalizer A device in the brake cable system that prevents one side from being applied before the other.

Floating caliper A disc brake caliper in which the caliper is mounted on pins or a slide to allow for some lateral movement.

Galled bearing Bearing surface damage caused by overheating, lack of lubrication, or improper lubrication.

Glycol A term used for ethylene glycol, an antifreeze solution.

Grease Lubricant containing a mixture of oil, soap thickeners, and other ingredients.

Heat A form of energy.

Heat checking Small surface cracks found in brake drums or brake rotors caused by excessive heat.

Hone To enlarge, smooth and clean a cylinder bore using an abrasive stone.

Hub The center part of a wheel, gear, or bearing.

Housekeeping A routine of cleaning and other practices to insure a safe and healthy workplace environment.

Hydraulic booster A type of brake power booster that uses hydraulic pressure from the power steering system pump.

Hydro boost A power brake booster that utilizes hydraulic pressure from the power steering system pump.

Hygroscopic The ability to absorb moisture.

Ignition switch A multi-position master switch, usually key operated, in a vehicle.

Integral ABS A self contained antilock brake system, which replaces the master cylinder and booster with a hydraulic modulator and high-pressure accumulator.

Lamp A device used to convert chemical energy into radiant energy, usually visible light.

Lateral runout A rotating member that has excessive variations in the amount of sideways wobble when turning.

Leaf spring A rear vehicle suspension system component consisting of one or more flat leaves of steel with graduated lengths.

Lining A friction material attached to a brake part used to slow or stop a vehicle.

Manifold A device used to hold two or more instruments for testing purposes; a device used to channel the air/fuel mixture into an engine; a device used to channel exhaust vapors out of the engine.

Micrometer A precision measuring tool.

Mineral spirits A cleaning solvent.

Modulator A device that regulates hydraulic pressure.

Nitrogen A high pressure odorless, colorless, tasteless gas that is often used to pressurize a system for leak testing.

Non-integral ABS An antilock system, which uses the existing, master cylinder and booster, but has ABS components added to the brake system.

Normally closed A term that refers to a switch or valve in its normal position, closed.

Normally open A term that refers to a switch or valve in its normal position, open.

Open circuit A circuit in which there is a break in continuity.

Pad wear indicator A mechanical or electrical warning device on the disc brake pad that warns of the need for pad replacement.

Parallelism Thickness variation in a brake rotor.

Pascal's Law When pressure is exerted on a confined liquid, the pressure is transmitted undiminished.

Piston A round caliper component in a disc brake; an aluminum or sintered iron component of a drum brake inside a wheel cylinder; the valve-like rod in a master cylinder.

Porosity A condition where fluids or gases can pass through the pores of a particular material.

Power steering A steering system that used hydraulic pressure to increase the torque applied to the steering wheel.

Pressure bleeder A device used to facilitate the removal of air from a brake system.

Pushrod A rod that transmits the movement and force of the wheel cylinder piston to the brake shoe.

Quick take-up master cylinder A master cylinder that supplies a large amount of fluid, under low pressure, during the first part of the brake application.

Relay An electro-mechanical switch.

Reluctor ring A wheel that triggers or pluses a magnetic field.

Residual pressure Remaining or leftover pressure.

Rotor A disc shaped brake component that rotates with the wheel.

Scan tool An electronic equipment device designed to communicate with vehicle on-board computer systems. Trouble codes and data can be extracted using a scan tool.

Shoe The lining and its steel backing of a drum brake.

Short circuit The intentional or unintentional grounding of an electrical circuit.

Silicone A group of organic compounds based on the element silicon (Si).

Sliding caliper A disc brake caliper that has piston(s) on one side of the caliper only.

Solenoid An electro-mechanical device used to impart a push-pull motion.

Specifications Technical data usually supplied by the vehicle manufacturer.

Spindle A shaft or axle on which a wheel hub or bearing rides.

Stoplight switch An electrical switch on the brake linkage or in the hydraulic system used to illuminate the brake lights when the brake pedal is depressed.

Torque steer An outside influence of a front-wheel drive vehicle, such as uneven front tire-tread wear that causes the steering wheel to turn right or left during hard acceleration.

Vacuum booster A power brake activation system that uses a vacuum signal on one side of a diaphragm to amplify braking effort.

Vacuum gauge A device used to measure vacuum.

Vacuum hose A small diameter rubber, plastic, or nylon tube used to transmit a vacuum signal.

Vacuum motor A diaphragm or motor-like device actuated by a vacuum.

Vapor A gas.

Warning light A light, usually on the dash, to warn of a problem.

Wear light The manufacturers' specifications as to the durability of a part in terms of serviceability.

Wheel sensor Magnetic speed sensor used to measure wheel speed in antilock brake systems.

Wiring harness The major assembly or a sub-assembly of a vehicle's wiring system.

Notes

Notes

Notes

Notes

Notes

Notes

Notes

Notes

Notes

Notes